FROM ANIMAL CHEMISTRY TO BIOCHEMISTRY

FROM ANIMAL CHEMISTRY TO BIOCHEMISTRY

NOEL G. COLEY, M.Sc., Ph.D.

Staff Tutor in the History of Science
The Open University

HULTON EDUCATIONAL PUBLICATIONS

ISBN 0 7175 0642 8

ACKNOWLEDGEMENTS

Acknowledgements are due to E. W. J. Bateman who prepared the Tables and to the following who kindly supplied the various photographs:

Plate 1 from Scientific Book Guild, Beaverbrook Newspapers Ltd., reprint of *Vegetable Staticks*, London, 1961.

Plate 2 from the Dover Edition (Constable) of Lavoisier's *Elements of Chemistry*, London, 1964.

Plates 3 and 4 by kind permission of the Science Museum.

Plates 5, 6, 7, 8, all from Wellcome Institute of the History of Medicine *except* No. 3. Plate 5 which is by kind permission of the Royal College of Physicians.

Plate 9 Radio Times Hulton Picture Library
Plate 10 The Mansell Collection
Plate 11 Liebig Museum, Giessen
⎫ prints supplied by the Open University Library.

Plate 12 by kind permission of the English Electric Co. Ltd.

Figures I-IV by kind permission of the British Museum.

First published 1973 by Hulton Educational Publications Ltd., Raans Road, Amersham, Bucks.

Printed in Great Britain by Northumberland Press Ltd., Gateshead.

Contents

Preface

WHAT DISTINGUISHES living matter from the inanimate; organised matter from organic compound? What gives rise to vitality? In fact, what is Life? Questions such as these have always been far easier to pose than to answer and attempts to find solutions to them have given rise to the highly sophisticated study of modern biochemistry. The object of this book is to trace the development of those parts of biochemistry which have grown from the study of animal matter and functions. This is only half the picture, of course, for biochemistry belongs as much to the chemistry of plants and a similar study is needed to trace the development of vegetable chemistry in order to obtain a balanced history of biochemistry in all its aspects.

Animal chemistry has a direct relevance for us which arises from its applications in medicine and the hope which it seemed to offer the nineteenth-century physician for a more scientific approach to diagnosis and treatment. For a long time this aim provided the necessary stimulus for the study of animal chemistry which consequently grew up within the disciplines of physiology and pathology. None realised more clearly than the physician the extent to which the animal functions rely upon the action of the nerves. Experimental evidence showed that the nervous influence resembled weak electrical impulses and the study of electrophysiology, one of the first branches of biophysics, began.

As the later researches of biochemistry have made clear, most physiological processes are extremely complex and involve long series of chemical changes, controlled by the presence of minute quantities of complex chemical substances such as enzymes, vitamins and hormones. These reactions and their sequences have proved exceedingly difficult to unravel even with the sophisticated methods and equipment of modern science. Hampered as he was by the inadequacy of contemporary chemistry, when faced with the problems posed by life-functions, it is surprising that the nineteenth-century animal chemist was able to proceed as far as he did in their solution and modern knowledge of biochemistry can only serve to enhance our admiration for his achievements. An interesting example of this is to be found in the Bence Jones proteins, discovered in 1845. These compounds remained a clinical curiosity until recent biochemical studies revealed that they are closely related in their chemical structures to the immunoglobulins. Complete amino-acid sequences for Bence Jones proteins have shown individual specificity, indicating that they are dependent upon the genetic information of the cells in which they are formed. Thus substances which were once regarded as degradation products of protein metabolism are now seen to be the result of biosynthesis and they have come to occupy a place at the very heart of modern biochemical research. Such exciting developments serve to emphasize the value of historical studies in science and if the present volume is found to place some of the more important biochemical problems in perspective its main purpose will have been served.

The author wishes to record his thanks to all those who have made helpful suggestions during the preparation of this work, but particularly to Dr W. H. Brock whose constructive criticisms have added so much to the coherence of the book. For all errors and omissions still remaining the author must take full responsibility.

Author's Note

Throughout this book common or trivial names for chemical compounds have been used so as to maintain as far as possible a consistent *historical* approach. To assist those who may be more familiar with the modern chemical nomenclature based on I.U.P.A.C. rules, a short list of the systematic names for the commoner chemical compounds is given at the back of the book. Further information about this system of nomenclature may be found in the booklet *Chemical Nomenclature Symbols and Terminology*, published by The Association for Science Education (1972).

Chapter 1

Early studies in the chemistry of Life

THE CAUSES OF LIFE have always aroused curiosity and from early times attempts have been made to provide some explanation of them. The functions of animal organs and the means by which food becomes changed into the substance of animal tissues remained completely incomprehensible for a very long time. It was commonly assumed that at some point in the course of changing inanimate food into living tissues within the animal body there must be the introduction of a vitalising principle. By this means it was thought that the change from mineral and organic substances present in the food to living or 'organised' tissue could be accounted for. But animals were seen to exhibit other puzzling phenomena such as the continuous production of heat; nervous irritability, muscular contraction, locomotion, the capacity to grow and so on. All these characteristics of living animals, together with others such as the heart beat and respiration, cease on death and the animal body thereafter becomes inanimate matter. What was it which left the body on death? In fact the living animal represents such a complex system that no simple answer to this question could be given and the attempts to provide one have led to the present sciences of biochemistry and biophysics. So far as can be seen at the present time these subjects are still very far from complete.

In the seventeenth century physical and chemical phenomena were generally explained in terms of particles of matter in motion—the so-called mechanical hypothesis. This was so successful that biologists were encouraged to adopt the same kind of mechanical approach to the study of the living organism. Perhaps the most notable outcome of this approach was due to William Harvey who in 1628 arrived at the concept of the circulation of the blood by applying quantitative measurement, calculation and deduction. Harvey regarded the heart as a pump, demonstrated the presence of valves in the veins and by measuring the capacity of the heart and the rapidity of its beats he calculated the total volume of blood expelled in a fixed period of time. He deduced from these observations that the heart continually pumps the *same* blood round the body in a one-way circulation, although he was unable to observe directly the capillaries connecting the arteries to the veins in the tissues. For this the microscope was required. Harvey was not himself a fully committed mechanist, but the success of his work encouraged others to suppose that life might be a purely mechanical process and that the animal body could be treated simply as a machine.

This was expressed most convincingly in the *Discourse on Method*, written by the French philosopher-mathematician, René Descartes and first published in 1637. Beginning with the circulation of the blood, Descartes accounted for animal functions in mechanical terms and showed that many of them were automatic.

'Nor will this seem in any way strange to those who, knowing the different kinds of automata, or moving machines, that the industry of men can fabricate with the use of a very few parts in comparison with the mass of bones, muscles, and so on, which are to be found in the body of every animal, will consider this body as a machine which, as it comes from the hands of God, is far better ordered, with far more wonderful movement than any machine that man can invent.'

This idea opened the way for the study of natural functions according to the laws of mechanics and it is not surprising to find that the method was particularly successful when applied to the functions of the skeleton. In 1680 Alphonse Borelli, an

Italian physicist, gave an account of the skeletal and muscular structures of the body in which the explanations were in purely mechanical terms.

Iatrochemists and the function of digestion

The mechanical approach had its limits however, for it could not be applied to the chemical processes of physiology. Nor did the mechanical theory give any account of the purposive behaviour of animals and the way they were adapted to their environment. In fact there was so much about the structure and functions of animals which was so complex as to be beyond comprehension that physiologists long felt the need to postulate a 'vital force' as the underlying cause of the phenomena of life. Both the living organism as a whole and its separate parts were often said to be imbued with souls or spirits and the organs thought to behave in an 'intelligent' manner so as to contribute appropriately to the balanced functioning of the whole. The attribution of rational behaviour to the parts of animals and sometimes to inanimate objects as well, has been called *animism*; it is an important ingredient in vitalistic theories. Thus, already in the seventeenth century, the foundations were laid for the argument between mechanist and vitalist which was to cast doubts upon the validity of all later attempts to explain life in physico-chemical terms. Controversies about vital force continued at least until the end of the nineteenth century. It always seemed possible to doubt whether chemistry alone could provide complete explanations for all the complexities of the life functions. Nevertheless, the need for the physician to find new ways of diagnosing, treating and preventing disease gave sufficient validity to the study of animal chemistry.

The study was first called iatrochemistry and its greatest exponent in the seventeenth century was J. B. van Helmont (1577-1644), who made some interesting and valuable discoveries in the chemistry of life. Details were often absent or misleading and some of the ideas which Helmont regarded as important were clearly animistic. He appeared to attribute powers of reasoning and purposive behaviour to purely chemical functions such as digestion and secretion, but he also recognised the acidity of the gastric fluid and suggested that it was neutralised by the alkaline bile in the duodenum. He thought that the food underwent fermentation in the stomach and was aware of the presence in that organ of something more than the mere acidity of the gastric fluid since he had found that vinegar alone would not digest bread. He therefore postulated that there was an 'archeus' in the stomach which directed the vital processes by the aid of ferments. The archeus was endowed with an understanding of the situation within the body and the ferments were more spiritual essences than material substances, but the fact that Helmont recognised the presence of hidden agents actively directing the life functions was in itself important. He regarded the digestive process primarily as a chemical change—a species of fermentation. The chyle formed from food as a result of the operations of the stomach and duodenum passed into the blood and was ultimately absorbed into the tissues. There was here then the rudiments of a chemical theory of digestion, beginning in the stomach and ending with absorption at the level of the tissues.

Sylvius de la Boë, professor of medicine at Leyden in Holland, attempted to give a more coherent explanation of physiological processes in purely chemical terms. He thought that digestion could be accounted for entirely by means of Helmont's acid and alkali, without the aid of the ferments. He also suggested that animal heat arose as a result of the chemical reaction between acid chyle and alkaline bile in which blood was formed in the heart. In addition to the bile, he recognised that saliva and pancreatic juice also played their parts in digestion, but he did not mention the gastric juice. Respiration was also regarded by Sylvius as a chemical process similar to combustion. He had noted that rapid combustion and rapid breathing both needed a copious supply of air, but he thought that the chief function of the inspired air was to cool the blood and restore the colour to arterial blood. Despite his predilection for chemical explanations, Sylvius could not entirely relinquish the notion of 'animal spirits' which he said were separated from the blood and rose up to the brain. This notion was derived from the physiological ideas of Galen, the Alexandrian physician who lived in the second century A.D.

Eighteenth-century studies in digestion

Although Helmont had declared the gastric juice to be acid, this was disputed during the seventeenth and eighteenth centuries. Most physiologists regarded the gastric juice as neutral or only faintly acid. Albrecht von Haller, the greatest German physiologist of the eighteenth century, thought that gastric juice was neutral and had no action as a 'ferment' but was simply a macerating liquid in which the food was softened and the process of assimilation begun. From the gastric juice, Haller suggested, the food received its vitality.

In 1752 Réaumur, the French natural historian best remembered for his thermometric scale, described some experiments with birds of prey including the kite which can regurgitate its food. By forcing these birds to swallow pieces of meat enclosed in small metal tubes fitted with wire gauze ends, Réaumur was able to demonstrate the solvent action of the gastric juice on food and also to estimate the rapidity of the chemical reactions during digestion. Sponges which the birds were made to swallow and then regurgitate, when squeezed out produced an opalescent fluid which turned blue vegetable colours red, showing the acid reaction of the gastric juice.

A few years later, in 1771, Edward Stevens a physician, repeated similar experiments on man and animals. He came to the interesting conclusion that digestion was not merely the result of heat, trituration and fermentation or putrefaction in the stomach alone, but was really caused by a powerful solvent secreted by the coat of the stomach. Stevens suggested that each species of animal probably secreted a specific variety of gastric juice capable of dissolving one type of food only.

Lazaro Spallanzani, professor of natural history at Modena in 1756 and later at Pavia, showed that gastric juice could digest food outside the stomach and so dispelled the idea that digestion could only occur in the *living* organ. He thought that the fluid from the stomach was not acid and that digestion was not a fermentation as some had suggested. Spallanzani had obtained his samples of gastric juice from a fasting stomach and this would account for their neutrality. His colleague, a chemist named Scopoli, made an analysis of gastric juice and found it to contain besides water, a fatty, gelatinous animal matter, *sal ammoniac* and earthy matter. He also noticed that gastric juice would precipitate silver nitrate solution due to the sal ammoniac (ammonium chloride), which it contained.

The commotion which could be observed during fermentation led some chemists to suggest that it was an intense motion amongst the insensible parts of a body, breaking down the old arrangements and giving rise to new. P. J. Macquer, professor of chemistry at the Jardin du Roi in Paris, expressed this view of the fermentation process about 1771. Like many other chemists of his day he distinguished three types of fermentation, viz., vinous, acetous and putrefactive, and he regarded these as three stages in the processes of decomposition. The release of fixed air (carbon dioxide) during fermentation and putrefaction led to the idea that this was the principle which held bodies together. It was thought to be due to the escape of this gas that the dissolution of the organic matter was brought about and it was often considered that the gases in the stomach and duodenum were beneficial in the digestive process.

Animal chemistry and the theory of phlogiston

Throughout the greater part of the eighteenth century theoretical chemistry was dominated by the phlogiston theory of Becher and Stahl. All combustible substances were thought to contain the principle of fire, or phlogiston, the source of heat and flames released during combustion. When the air became saturated with this principle the fire was extinguished. All our modern ideas about elements and compounds were inverted in phlogistic chemistry because all combustible substances whatever were held to contain phlogiston. Thus elements such as carbon, hydrogen, sulphur and phosphorus, together with all the metals must contain this principle, but carbon dioxide, water and the oxides in general do not. It followed that the products of combustion must be considered to be simpler than the combustible substances from which they were derived.

Modern historiography has shown that the phlogiston theory did not seriously hamper developments in inorganic and mineral chemistry until the discoveries of the eighteenth century pneumatic chemists began to increase the numbers of known identifiable gases and the use of quantitative methods indicated that the products of combustion weighed more than the elements alone. Nevertheless, phlogistic chemistry was quite incapable of dealing effectively with the complexities of animal chemistry. The main method of animal analysis was by destructive distillation in which the animal substance was heated in the absence of air. As a result of this procedure a small number of watery and oily 'spirits' were generally obtained, together with some 'air' and a residue of charcoal. These products were often thought of as the 'proximate principles' from which the animal matter was made up. They were thought to pre-exist in it and were merely released from it by the action of heat. These spirits were also said to contain different proportions of phlogiston and thus the true chemical composition of the animal matter was completely obscured.

The problem of animal heat

Perhaps the most remarkable characteristic of living things, distinguishing them from the inanimate, is their ability to maintain a constant temperature. Amongst the warm-blooded animals this is usually well above that of their surroundings. In this respect animals disobey Newton's law of cooling and seem almost to contravene the second law of thermodynamics. The problem of the origin of animal heat had occupied the attention of chemists and physiologists since the early years of the eighteenth century and had long been at the centre of the controversy between mechanists and vitalists. The most popular theories about the origins of animal heat in the eighteenth century were mechanical and depended upon friction between the circulating blood and the walls of the blood vessels, or between the corpuscles of the blood itself.

Hermann Boerhaave, the great Dutch chemist, for example suggested that the blood acquired most of its heat as a result of friction in its passage through the lungs. The purpose of respiration, he thought, was merely to cool the blood, since the additional heat which is gained in the body from the production of chyle would otherwise make it too hot. Stephen Hales, vicar of Teddington and famous for his attempts to apply Newtonian principles to plant and animal functions, expressed a similar view and even went on to calculate the degree of refrigeration produced by respiration in the lungs. Hales thought that friction in the lungs was responsible not only for the greater part of the heat in the blood but also for its florid colour. It appeared that the functions which the lungs were thought to maintain were both complex and contradictory. The major weakness of any theory in which animal heat was thought to be generated entirely in one organ was that it then became difficult to explain why the whole body should exhibit a uniform temperature if most of the heat were produced at one point.

Despite its limitations however, the friction theory of animal heat was adopted by Haller, who relied on common sense arguments taken from observations in comparative anatomy and physiology to supply evidence for the theory. Thus, fishes with a small heart and feeble circulation are cold-blooded, but birds with comparatively large hearts and rapid circulation, have a high body temperature. In man too the temperature rises with violent exercise and this might well be due to the increased circulation, Haller thought. He would not allow that chemical reactions could account for animal heat, nor would he admit that vital force played any part. His outlook was therefore entirely mechanistic and the influence of his work, which was very wide, helped to ensure that by the mid-eighteenth century there was a general acceptance of the mechanical theory of animal heat amongst physiologists.

Heat exchange processes to account for animal heat

Animal functions seemed to many to resemble putrefactive and fermentative processes which were known to require air and to yield warmth. In the animal body such processes were thought to occur to the greatest extent in the lungs, where more heat was produced the faster the blood circulated. Considered in this way the production of animal heat became a chemical process. Joseph Black, the Edinburgh chemist famous for his discovery of latent and specific heats about 1760, considered that the main source of animal heat was to be found in the process of respiration. Daniel Rutherford, a colleague of Black, was influenced by this theory and suggested that animal heat and the alteration of the air which accompanied breathing were due to the same cause. Now it was well-known that the air became 'phlogisticated' during respiration and so it seemed that the phlogiston liberated by the animal might account for its elevated temperature. Phlogiston was thought to be evolved from the blood in all parts of the body and the function of respiration was to remove the phlogiston by dissolving it, whilst at the same time the blood was cooled. One merit of this theory was that it overcame the objection that all the heat was liberated in the lungs by suggesting that heat was released from the blood throughout the body.

Adair Crawford, physician at St Thomas's Hospital in London, visited Scotland and became interested in Black's experiments on heat. He therefore set out to try to establish the heat-exchange theory by supplying experimental data to support it. He began by finding the specific heats of a number of substances including blood, using the method of mixtures. Blood was found to have a high heat content which Crawford suggested came from the air during respiration. It was known that animals with large lungs were warmer than those with small lungs and slower respiration. This appeared to provide qualitative support for Crawford's theory, the main points of which were:

1. A given volume of atmospheric air contains more heat than the same volume of air after expiration from the lungs. Crawford thought that he could demonstrate this experimentally and he gave the ratio of 'absolute heat' in dephlogisticated air to that in atmospheric air as 4.6:1.
2. Arterial blood contains a higher proportion of heat than venous blood. This too Crawford demonstrated, giving a ratio of 'absolute heat' in arterial blood to that in venous blood of 11.5:10.
3. Heat and phlogiston are complementary and the addition of phlogiston to a body reduces its heat capacity (and *vice versa*).

During the circulation of arterial blood then, heat would be evolved as phlogiston was absorbed. This phlogiston would be given off in the lungs where it would be exchanged once more for heat from the air. In this way heat was thought to be evolved from the blood in all parts of the body in proportion to its degree of phlogistication. When there was an increase in the rate of circulation, as during exercise, the evolution of heat throughout the body would also become more rapid. In Crawford's account of the formation of animal heat, as in combustion, the origin of the matter of heat was the air. In respiration this heat was first taken up by the blood in the lungs and later released gradually throughout the body. Crawford was widely credited by his contemporaries with having devised a new theory of animal heat.

Apart from the very real difficulties of confirming Crawford's experimental results for the heat contents of air and blood, his theory could be criticised on the grounds that purely physical explanations were inadmissible when dealing with the vital functions. The alternative to all such mechanical, chemical, or physical explanations was to consider the production of animal heat as a function of the vital force setting itself against the processes leading to decay by physical and chemical changes. The power of maintaining a constant temperature above that of its surroundings, observable in any living organism, might well be due to some principle peculiar to life itself. This was the position adopted by John Hunter, the leading English physiologist of the eighteenth century, and by other physiologists who accepted the vitalistic conception of life.

Distillation analysis

Stephen Hales in his *Vegetable Staticks*, published in 1727, gave detailed descriptions of experiments in which he extracted and measured the volumes of 'air' contained in various plant and animal substances. His apparatus consisted of a wide-necked globe luted onto the neck of a retort which was heated in a furnace. The gases evolved during the distillation were collected by displacement of water from the globe and their volumes were measured when the whole apparatus had cooled to room temperature. Hales found that he could obtain large quantities of gases by heating blood, fat and even the more solid parts of animals in the absence of atmospheric air. Hog's blood, tallow and deer's horn all produced a large volume of 'air' together with white fumes of sal ammoniac or 'volatile salt'. Other substances which yielded large quantities of gases included oyster shells, honey and beeswax, but by far the greatest volume was evolved from a urinary calculus. Hales knew that Rhenish tartar (probably potassium tartrate), gave off a large volume of 'air' when heated and he observed that this behaved similarly to the gas from the calculus in that they both dissolved when confined over water. From this he wrongly concluded that urinary calculi were composed of 'a true animal tartar' and he likened the formation of such stones in the animal body to the deposition of tartar in old wine-casks. In this way Hales showed that he accepted the ancient tartar theory of disease which had been revived and re-introduced into medicine by Paracelsus in the fifteenth century. According to this theory minute quantities of poisons introduced into the body with the food, accumulate as deposits on the teeth and in the organs, where they cause diseases such as gout and stone.

Hales work was the immediate forerunner of the discovery of 'fixed air' (carbon dioxide) in its role as a chemical constituent of solids by Joseph Black in 1756. The notion that the air contained in solid bodies served to hold them together was often proposed in the eighteenth century. Hales suggested that the attracting powers of the inelastic air particles served to form anomalous concretions in animals. The influence on Hales of Newton's gravitational theory is apparent here. This idea was also supported by David Macbride who investigated the role of fixed air in the processes of digestion, fermentation and putrefaction.

C. W. Scheele and the chemistry of natural products

The Swedish apothecary C. W. Scheele carried out a variety
of experiments on natural substances of both plant and animal
origin. His experimental accounts were outstanding for their
care and thoroughness. One of Scheele's most significant contributions to this work was his recognition of the need to seek
the highest possible degree of purity. This is clearly necessary
for success in any attempt to analyse organised matter, but it is
very difficult to achieve with animal substances and the problem
of purity was to remain a limiting factor in animal chemistry
throughout the nineteenth century.

Scheele's first chemical experiments on organic substances
were concerned with the acids. He obtained pure samples of
tartaric, citric and malic acids by adding dilute sulphuric acid
to solutions of their calcium salts. He also prepared mucic and
oxalic acids by oxidation of milk sugar, whilst pyromucic acid
was prepared by heating mucic acid and lactic acid was
obtained from sour milk. These preparations were important
in animal chemistry because they were seen to yield recognisable organic compounds from organised matter or its derivatives. Scheele also carried out an important investigation into
the nature of urinary calculi. He was the first chemist to recognise uric acid—although he did not give the substance its name.
The red coloration produced when uric acid is heated with nitric
acid was known to Scheele who used the reaction as a means
of identifying uric acid. He also obtained a sublimate on heating uric acid which he thought was succinic acid (it is in fact
cyanuric acid). Both Scheele and his teacher Tobern Bergman,
thought that *all* urinary calculi contained mainly uric acid,
although A. S. Margraaf the German chemist, had already in
1746 obtained phosphates from the urine and it would seem
reasonable to have assumed that these salts too must sometimes
be deposited and appear in urinary calculi.

Scheele's phlogistic views led him to suppose that the preparation of organic substances was often merely a matter of analysis. Thus he treated alcohols with acids to obtain esters which he called 'ethers' and which he thought pre-existed in the alcohols themselves. He also considered that the products of combustion of oils were really constituents of the oils themselves with phlogiston. Thus,

Oil = Fixed air + Water + Phlogiston.

In fact, as we have seen, the phlogistic chemists considered any combustible substance as a compound containing phlogiston and the process of combustion merely released the constituents of the combustible substance. In the same way the process of destructive distillation, or heating in the absence of air, was thought to lead to the release from organised matter of its pre-existing components or 'proximate principles'. Such conceptions could only lead to confusion and in fact until elementary organic analysis was introduced in the nineteenth century no clear idea of the true composition of organic and organised substances was possible.

Important both for its quantity and variety as well as for the attention to detail which it displays, Scheele's chemical work marked a distinct advance on anything which had gone before. It was significant for the animal chemist because it demonstrated the fact that natural substances derived from living organisms could be treated by the same methods as were being successfully applied in mineral chemistry. Natural substances were seen to be related to each other in ways which were similar to those found amongst the well-known inorganic compounds. It is in the work of Scheele that we see the true beginnings of animal chemistry. He thought that it would be possible to discover the elements of which organic matter is composed and although Scheele's 'elements' were no more than 'proximate principles', this idea was also valuable. It gave support to the application of chemical analysis to living beings and their products, and encouraged the animal chemist to seek out the chemical composition of such substances. Nevertheless, little of real value could be achieved in the complex study of animal products until the fundamental concepts of chemistry had been clarified and defined.

Chapter 2
Founders of Animal Chemistry

Lavoisier; respiration and animal heat

IT IS IN THE WORK of Antoine Lavoisier that the origins of all later developments in chemistry can be found. In addition to his well-known experiments leading to the establishment of the oxygen theory and the overthrow of the concept of phlogiston, Lavoisier provided the chemist with a viable definition of the chemical element as the last point in analysis. He showed that organic substances are composed of a small number of common elements and indicated the method by which the proportions of these elements in organic compounds might be determined. Lavoisier insisted upon a systematic quantitative approach to experimental chemistry and as a result of his work he was able to demonstrate conclusively the part played by oxygen in respiration as well as in combustion and calcination. In his famous 'ideal experiment' on the calcination of mercury, Lavoisier was able to show that the air is composed: of two parts which he called respirable air and mofette. The description of this experiment was given in a paper published in 1780 in which Priestley was mentioned several times as having taught that respiration resulted in the phlogistication of the air, as in the calcination of metals. Respiration was thus recognised as analogous to calcination in its effects.

Lavoisier found that when a sparrow was made to breathe a confined volume of air until it died the residual air would extinguish a candle and turn lime-water milky. The decrease in volume was not more than one-sixtieth part, but on adding caustic alkali a reduction of one-sixth occurred and the alkali was rendered 'mild', showing that fixed air was present. The residual mofette (atmospheric nitrogen) was found to be similar to that remaining after the calcination of mercury. Thus, in respiration as in combustion, respirable or 'true air' was removed and was replaced by fixed air, whilst the mofette was to be regarded as a purely passive medium which entered and left the lungs without apparent change. This came to be doubted later, particularly in the case of herbivores which were sometimes thought to derive some of their nitrogen from respired air.

Two explanations for the process of respiration seemed to be possible. The respirable air might either combine with the blood, or be changed into fixed air. Both processes were thought to occur at the same time, part of the respirable air combining with the blood and part being changed into fixed air. By analogy with the red colour of certain metallic oxides, Lavoisier suggested that the respirable air acted upon metallic elements in the blood and so produced its red colour. He soon abandoned this comparison with metallic oxides, but a return to it was made when Berzelius and others in the nineteenth century were able to confirm the presence of iron in the blood.

In 1777 Lavoisier wrote an important general account of combustion in which he proposed the main points of his oxygen theory. This, he said, explained satisfactorily the phenomena of combustion, calcination and the respiration of animals. Combustion, which could only occur in 'pure air' (i.e. oxygen), was accompanied by the disengagement of the matter of fire and light. The pure air was destroyed and the burnt body was found to increase in weight in exact proportion to the quantity of pure air destroyed. The changes produced in the burnt substance resulted in the formation of an acid. Calcination was regarded by Lavoisier as a slow form of combustion, whilst in respiration he said, '. . . the pure air passing through the lung undergoes a decomposition similar to that occurring in the combustion of carbon.'

At this time Lavoisier was merely proposing these points as an alternative hypothesis to that of phlogiston. One of the chief differences in the new theory was that the matter of fire was to be found combined with the base of pure air or oxygen, instead of being considered as a constituent of the combustible substance. According to Lavoisier, during the change from pure air to fixed air the matter of heat or 'caloric' was disengaged from the former. This was thought to be the sole source of animal heat and since the change occurred only in the lungs, all the heat of the body must necessarily be produced in these organs. It was then distributed throughout the body by means of the circulation of the blood. This theory was a retrograde step from that of Crawford.

In 1780 Lavoisier wrote a famous Memoir on heat in collaboration with the French mathematician Laplace. Their experiments, based on the use of the well-known ice calorimeter, laid the foundations of thermochemistry. They included determinations of the specific heats of a great variety of substances and the heats of reaction in a number of chemical changes, especially combustion reactions. The most significant of these determinations from the point of view of the animal chemist were the attempts to measure the heat of combustion of carbon and to correlate this with determinations of animal heat. Whereas Crawford, as we have seen, regarded the evolution of animal heat as a purely physical process resulting from the redistribution of the matter of heat due to changes in the specific heats of venous and arterial blood, Lavoisier regarded the production of animal heat as a chemical process analogous to combustion.

Lavoisier and Laplace made experiments on the respiration of birds and guinea-pigs, observing that in pure air (oxygen) the sole product of respiration was carbon dioxide. Regarding this change as the main source of animal heat, they measured both the heat of combustion of carbon and the quantity of carbon dioxide formed in a given time by the respiration of a guinea-pig. From this they were able to calculate the heat evolved by the animal in ten hours, assuming it all to be produced as a result of this chemical reaction. They later placed the animal in the ice calorimeter and measured the weight of ice which was melted directly. Reasonable agreement was obtained between the results for the heat evolved by the animal in the two parts of the experiment, but since the value for the heat of combustion of carbon used by Lavoisier and Laplace was very inaccurate, the agreement was quite fortuitous. In fact it seems that their results were obtained largely by the chance cancellation of errors and later experimenters, using the same apparatus, failed to obtain consistent results. Nevertheless, this work established the basis for a true theory of respiration as a slow form of combustion.

Lavoisier made the mistake of supposing that all the animal heat originated in the lungs (neither he nor Laplace had much knowledge of physiology), and that it was distributed from this organ by the rapid circulation of the blood. They also suggested that the heat balance in the body was maintained partly by the aid of the cooling effect of the air in the lungs due to evaporation and partly by changes in the specific heat of the blood as Crawford had said. The volume of oxygen used up during exertion and digestion was found to be greater than that consumed during periods of rest and the increased oxygen intake was found to be accompanied by an increase in the temperature of the body. These observations showed that respiration, nutrition and other vital activities of the body were all linked and since respiration could be explained chemically it seemed likely that the other animal functions were also chemical. Indeed, Lavoisier's successful theory of respiration as a chemical process occurring in the lungs was more than a simple improvement on the earlier theories—it established a precedent for animal chemistry since it showed that the chemical approach to life processes clearly worked.

In his *Traité Élémentaire de Chimie*, published in 1789, Lavoisier introduced a new definition of the chemical element, based on practical considerations of chemical analysis. This definition came to be accepted by virtually all chemists and it provided the theoretical basis for the methods of elementary organic analysis developed during the nineteenth century.

> '... if we apply the term *elements*, or *principles of bodies*, to express our idea of the last point which analysis is capable of reaching,' wrote Lavoisier, 'we must admit, as elements, all the substances into which we are capable, by any means, to reduce bodies by decomposition.'

Lavoisier found that animal substances were in general composed of a few such elements only, chiefly carbon, hydrogen and azote (i.e. nitrogen), with phosphorus and sulphur appearing occasionally. He rejected the view, then commonly held, that the products of combustion or destructive distillation pre-existed in the organised matter and his elements were so defined that chemists could seek them in *all* matter. In organic compounds the elements were generally thought by Lavoisier to occur in groups or 'radicals' which combined with oxygen. Organic acids in particular were thought to be oxides of such radicals. In sugars the hydrogen and oxygen was found to be in the proportions 2:1, as in water, and many animal substances were known to contain nitrogen.

Lavoisier also studied the processes of fermentation and putrefaction which were commonly thought to be involved in natural functions such as digestion. In his explanation of the mechanism of vinous fermentation, Lavoisier assumed that water was first decomposed into its elements (by 1787 he had determined the true composition of water as an oxide of hydrogen). During fermentation the oxygen combined with carbon in the sugar to form carbonic acid which was given off, whilst the hydrogen of the water combined with another part of the carbon to form alcohol. Since he accepted the principle of conservation of matter, he also held that if it were possible to re-unite these two products, sugar would once more be obtained. 'We may consider the substances submitted to fermentation and the products resulting from that operation as forming an algebraic equation', he wrote. Thus,

Grape juice=carbonic acid+alcohol.

In his discussion of this process Lavoisier came nearer to the modern view of organic constitution than in other parts of his chemical theories. He suggested that in sugar the elements were so delicately combined together that very gentle forces were all that was needed to destroy their equilibrium. This was brought about in practice by adding one-tenth of the weight of yeast to a weighed quantity of sugar and Lavoisier devised an apparatus in which this could be done quantitatively. His latest modification of this apparatus is still to be seen in Paris, but it is likely that Lavoisier never used it himself because he was put to death during the course of his experiments on this subject.

Lavoisier also regarded putrefaction as a form of fermentation, as did Stahl. In this case some of the hydrogen was released as a gas, much of the carbon was found to be oxidised to carbonic acid, whilst the azote, sulphur and phosphorus were given off as ammonia, hydrogen sulphide and phosphine respectively. In the end nothing remained except a small proportion of earthy matter mixed with some carbon and iron. In this way putrefaction could be represented as an analysis of the organic material. Lavoisier was of the opinion that nothing could change the course of the putrefactive process. He regarded both fermentation and putrefaction as chemical processes which obeyed the ordinary rules of chemistry. The acetous fermentation was slightly different in that it was an oxidation process requiring the presence of oxygen or air, but it was nevertheless a chemical reaction. This view was disputed by those who thought that all animal and vegetable processes were 'vital' and could only proceed in the presence of the vital force.

About 1789 Lavoisier began another series of experiments

on respiration in collaboration with Armand Séguin, an army contractor. Measured volumes of vital air (oxygen), atmospheric air or other respirable gas mixtures were breathed under controlled conditions by human subjects using a face mask. From their results Lavoisier and Séguin concluded that respiration is a true combustion of both carbon and hydrogen in the blood, comparable with the burning of a candle flame. They regarded atmospheric air as a compound of vital air and azotic gas (nitrogen), though it must be remembered that the distinction between a chemical compound and a physical mixture had not then been understood. During respiration the air was thought to supply oxygen and caloric whilst the blood supplied the combustible substances—carbon and hydrogen— which are constantly being renewed from the food. Hence respiration could be linked with nutrition in the lungs and the interaction of these two processes would result in the evolution of animal heat.

Lavoisier and Séguin found that the quantity of oxygen used up in respiration was the same whether pure vital air or vital air mixed with a high proportion of azote were used and there was neither absorption nor evolution of azote during the process. The quantity of oxygen consumed was found to rise with temperature, during digestion and as a result of movement or exercise. In all these cases however, the temperature of the blood remained almost constant. They were also able to show that respiration proceeded equally well in a mixture of vital air and hydrogen as in atmospheric air. The total quantity of oxygen consumed by a man during 24 hours was calculated and the volumes of carbon dioxide and water vapour formed from it were found.

Lavoisier also compared the water vapour exhaled from the lungs with that given off from the skin by transpiration. He and Séguin thought that the carbon dioxide exhaled was actually *formed* in the lungs, but they admitted the possibility that some carbon dioxide might have been formed during digestion and then transferred to the circulation in the chyle. In this case it would merely be *released* from the blood when it arrived at the lungs, the carbon dioxide being replaced by oxygen. Though the numerical results given by Lavoisier and Séguin were not very accurate, the methods used were important since they pointed the way for other workers in the same field throughout the nineteenth century. Unfortunately, before these experiments could be completed, Lavoisier was to meet his untimely end.

Fourcroy and Vauquelin

If Lavoisier had shown the way in which chemistry might profitably be developed, he was ably supported by others of his fellow countrymen. Amongst those who took up the study of animal and vegetable chemistry the most successful and best known was Antoine François de Fourcroy (1755-1809). Born in poor circumstances, Fourcroy was for a long time in no danger from the zeal of the French Revolutionaries. He received a good grounding in chemistry and in 1784, on the death of P. J. Macquer, he was appointed to the professorship of chemistry at the Jardin du Roi in Paris (later to become the Muséum National d'Histoire Naturelle). Fourcroy is said to have been a fine lecturer and certainly his popularity in this respect continued to increase throughout his life. His eloquence is reflected in his books. The chemistry of animal substances formed his principal interest and his work in this field added a wealth of much needed facts.

Fourcroy welcomed the anti-phlogistic theory from its inception. In 1782 he published a book based on his lectures in which he indicated that he had compared the phlogistic and anti-phlogistic theories. Whilst he was somewhat undecided between the two, he was attracted to the new theory because it was founded upon experimental data and was therefore more convincing. The second edition of this book appeared in

1786 and it has been suggested that during the preceding two years Fourcroy was finally persuaded to accept the new theory without reservation. Yet in the text of this second edition the two theories are still given as alternatives. Thus, although he was attracted to Lavoisier's theory, Fourcroy was clearly not completely convinced in its favour and this is reflected in the approach he adopted to 'animal analysis' which he based on 'proximate analyses' and qualitative tests to identify animal substances.

It was in his great textbook, *Système des Connaissances Chimiques*, first published in 1801-2, that Fourcroy outlined his views on animal chemistry. The division of organic chemistry into animal and vegetable chemistry which Fourcroy adopted in this book was to govern the organization of the subject during the whole of the nineteenth century. Among the earliest work on animal chemistry carried out by Fourcroy was an examination of urinary calculi and an analysis of the urine. In this work he was assisted by L. N. Vauquelin, who is said to have performed most of the experiments.

Vauquelin had become Fourcroy's laboratory assistant in 1780. He was industrious, simple and honest, though Humphry Davy rather snobbishly considered that his methods were those of a pharmacist (i.e. a *tradesman*), rather than a *philosophical* chemist. This may have been true, but Thomas Thomson said that Vauquelin was by far the most industrious of the French chemists and that he, '... published more papers consisting of mineral, vegetable and animal analyses, than any other chemist without exception.' Fourcroy and Vauquelin jointly published papers on animal chemistry from about 1790 onwards and in addition Vauquelin published many analyses of animal substances in his own name. Tears, nasal mucus, hair, synovial fluid (the fluid which lubricates the joints), the amniotic fluid of the foetal calf, uric acid, bones, etc., were all examined, as were milk, blood, bile, urine and other animal fluids.

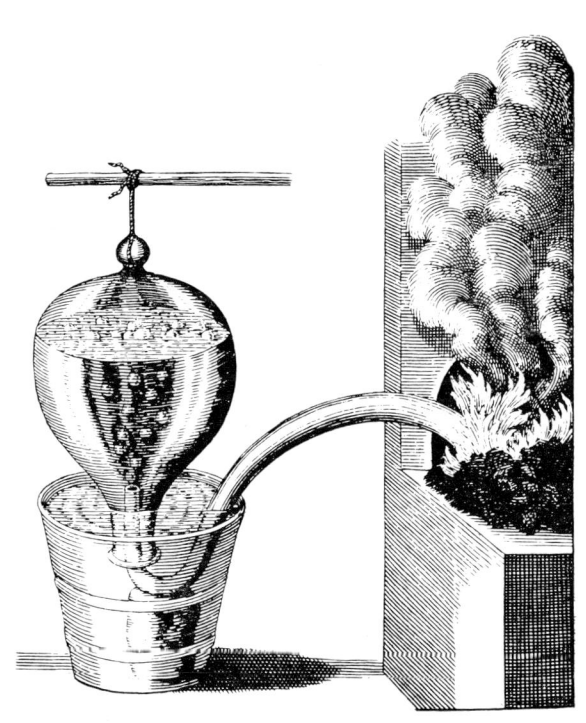

Plate 1.   Stephen Hales' Distillation Apparatus.
(From S. Hales, 'Vegetable Staticks', 1727)

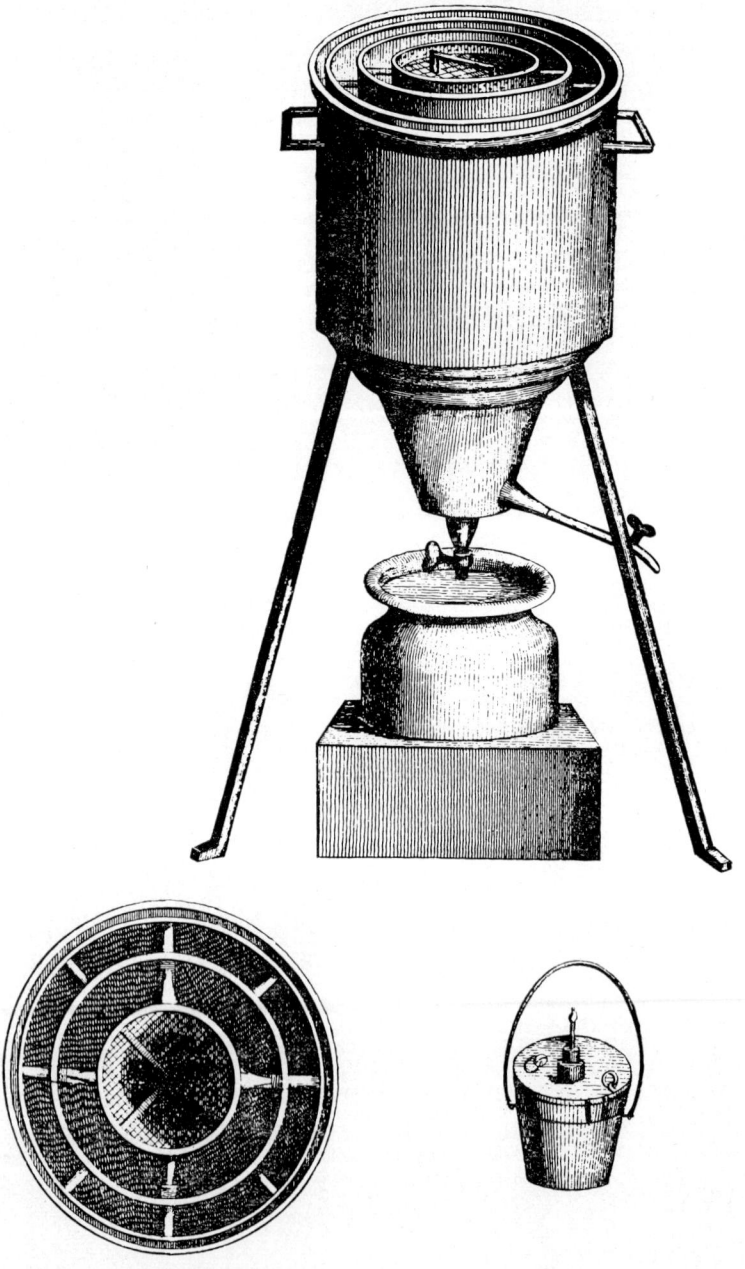

Plate 2. The Ice Calorimeter of Lavoisier and Laplace (*ca.* 1780).
(From A. Lavoisier, 'Traité Élémentaire de Chimie', 1789, trans, R. Kerr, Edin., 1790)

In 1799 a joint paper on the composition and chemical properties of the urine appeared. Fourcroy and Vauquelin treated this topic in a similar manner to that used in examining mineral waters. They found a number of mineral salts in the urine in addition to organic matter. These included the muriates (chlorides) of sodium and ammonium and the phosphates of calcium, magnesium and ammonium, both singly and combined as double salts. They also found uric acid and another acid which they thought to be benzoic. In addition there was a small proportion of gelatin and albumen which varied with the health of the subject. Most important they recognised urea, the substance which was considered to give the urine its peculiar properties. Fourcroy and Vauquelin isolated a fairly pure sample of urea in the form of slightly discoloured white crystals.

The acid which they thought to be benzoic acid was obtained by sublimation from the solid residue of the urine after evaporation to dryness, or by adding concentrated hydrochloric acid to the concentrated solution produced by evaporating urine to a syrupy consistency. A similar acid could be obtained in quantity from the urine of the horse and cow, but there was some doubt as to the identity of these two acids. Berzelius was unable to find benzoic acid in human urine and others showed that when benzoic acid was included in the diet it was excreted as hippuric acid. It was Liebig who finally pointed out that benzoic and hippuric acids had frequently been confused and that benzoic acid never appeared in the fresh urine of a healthy human subject, although minute quantities of hippuric acid were usually present. This confusion, difficult or even impossible to avoid in the absence of adequate tests, illustrates some of the problems of identification facing the animal chemists. The chemical compositions of these two acids are fairly similar, although hippuric acid ($C_6H_5CO.NH.CH_2COOH$), contains a few more atoms per molecule than benzoic (C_6H_5COOH). The important difference from the point of view of animal chemistry is that the first contains nitrogen whilst the second does not, and so to confuse one for the other must lead to a mistaken idea of metabolism.

The presence of nitrogen in animal matter caused early animal chemists some conceptual difficulties. It was commonly thought that the process of 'animalisation' which occurred as food was assimilated into the body tissues, was brought about largely by the increase in nitrogen content together with changes in the proportions of other elements. The source of the additional nitrogen in the carnivore was obvious enough, but it was difficult to see how a herbivore obtained all its extra nitrogen since the proportions of this element found in green plants were always small. However, in 1789 Fourcroy had shown that plants often contained both albuminous matter and gelatin and he was thus able to confirm the discovery made by Berthollet that some forms of vegetable matter contain nitrogen. When he later found that the gluten of flour contains nitrogen, he concluded that this might well be one important source of that element in the body. Nitrogen was found in the swim-bladder of the carp by Vicq d'Azyr who thought that it was produced by decaying food in the stomach. Fourcroy accepted this explanation and suggested that nitrogen might be one of the products in the process of putrefaction.

In 1790 Fourcroy made some macabre observations which supported this conclusion when he examined the bodies in the crowded cemetery of the Innocents in Paris. For health reasons this ancient burial ground was to be cleared and it was found that in the multiple graves, some of which contained up to fifteen hundred bodies, the muscular tissues had changed into a kind of fat which Fourcroy called *adipocire* and which resembled spermaceti, a waxy substance found in the oil of the sperm whale. A few years earlier he had examined a specimen of human liver which had been exposed to the atmosphere for about ten years until all putrefactive processes had ceased. He had found similar properties in this material to those of adipocire. When exposed to the air, this substance dried off but did not diminish in volume; it was whitened but had lost its characteristic smell. On heating the material, quantities of ammonia were evolved. Fourcroy also examined the behaviour of this substance with water, alkalis, acids and alcohol. His results led him to conclude that during putrefaction animal substances lost their nitrogen and were converted into a kind of waxy or soapy matter.

About this time too, Fourcroy also examined the matter of biliary calculi and again found a similar substance to that of the putrefied liver. He suggested that it might be a function of the liver to excrete this substance from the body; gall stones could then be explained as an excess of it. Fourcroy determined the solubilities of the three forms of fat in alcohol and also found their melting-points. He regarded these fats as three varieties of a single substance—adipocire. His numerical results were important at a time when so few quantitative measurements were available in animal chemistry.

Gall-stone derivative	Product of putrefaction	Spermaceti
Hot alcohol dissolves one-ninth of its own weight of this fat. M.pt. 90°R.*	Hot alcohol dissolves twice its own weight of this fat. M.pt. 28-30°R.	Hot alcohol dissolves just over its own wt. of this fat M.pt. 32-34°R.

* Temperatures measured on the Réaumur scale.

TABLE I

This work inspired Chevreul to undertake the investigation of the fats in the early years of the nineteenth century.

Experiments on the composition of the blood led Fourcroy to conclude that muscle fibre is produced from the blood by means of a vital process which could only occur in the living body. When he examined the blood of individuals of different ages he found that the proportion of fibrous matter increased with age as also did the amount of fibrous matter in the muscular tissues. Blood left in contact with vital air (oxygen) for eight days was found to absorb some of the gas with change of colour and at the same time fixed air was evolved. Changes of this kind had already been observed by Richard Lower, Robert Hooke and other founders of the Royal Society in the seventeenth century. They indicated the part played by the blood in the process of respiration, but Fourcroy did not pursue this line of investigation; he returned instead to the examination of the blood by distillation, coagulation and its behaviour towards heat and water. He extended his experiments to the blood of the human foetus which he found to be devoid of fibrinous matter and incapable of clotting when cooled.

In his study of the processes of digestion Fourcroy displayed a particular interest in the function of the bile. He thought that the soda in this fluid served to neutralize the acid from the stomach, whilst the fatty portion combined with undigested food and so separated it from the digested part which then formed chyle. The latter was absorbed into the blood stream where it began its conversion into blood, a process which was completed in the lungs where the inspired oxygen carried off the excess carbon and hydrogen. Theorising of this kind was valuable for the animal chemist because it gave him a framework within which to fit his observations. It was from the same starting point that William Prout began his studies and later in the nineteenth century Liebig was also to build his controversial scheme of animal chemistry upon experiments and speculations concerning metabolism.

Fourcroy examined the brain matter of calves, sheep and man. He found phosphates of lime, soda and ammonia, together with organic substances which he found to be different from any previously known. Later, in 1812, Vauquelin made a thorough investigation of the composition of the brain which was not surpassed until the work of Thudichum, seventy years later. Vauquelin found two fatty substances in the brain, together with albumen, 'osmazome', a number of salts including the phosphates of potash, lime and magnesia, phosphorus and sulphur. He gave the following proportions,

water	80.00
white fatty matter	4.53
reddish fatty matter	0.70
albumen	7.00
osmazome	1.12
phosphorus	1.50
acids, salts, sulphur	5.15
	100.00

TABLE II. *Vauquelin's chemical analysis of the brain*

(Such analyses were always adjusted by the introduction of general categories and 'rounding off' the figures so that they added up to exactly 100 per cent.)

In their analysis of the soft roe of the carp, Fourcroy and Vauquelin again found phosphorus which, they suggested, was in direct combination with carbon because they could not extract phosphates from the original roe, although these salts were to be found in the residue left after calcination. They suggested in passing that this phosphorus was responsible for the phenomenon of phosphorescence in rotting fish. The recognition that some of the processes of analysis then in use resulted in chemical changes in the substances under examination was important because it emphasised the need for more delicate ways of testing these complex animal materials.

The chemical work of Fourcroy and Vauquelin is full of interesting observations about animal substances. For example, in their analysis of milk published in 1806, they made the observation that the phosphates of lime and magnesia are contained in this fluid in the ratio of 50:1 as in bones. From this they concluded that these substances in milk gave rise to bone formation in young animals. At the same time, iron was absorbed into the blood and the curd of the milk went to form muscle tissue, they suggested. Some forms of mucus were found to dry to a hard mass almost insoluble in water and they thought that such a substance might well form the epidermis of certain land animals. Their analyses showed that hair, wool, silk and feathers were all chemically similar and were related to skin, although there were possibly differences in the relative proportions of calcium and magnesium phosphates present in each of these substances. Another interesting discovery made by Fourcroy and Vauquelin was that of the presence of fluoric acid in dental ivory. They found that the proportions of this compound increased with age and that there is a much higher percentage of it in fossil ivory. To account for this remarkable observation they suggested that the fluoric acid was slowly absorbed from the surroundings with the passage of time. This has been found to be the correct explanation of the presence of fluorides in fossils and in fact the age of certain fossils may be estimated from the fluorine content.

Berzelius and the analysis of animal fluids

From about 1806 onwards, Berzelius in Sweden became interested in the problems of physiological chemistry mainly because he saw the relevance of this study to the practice of medicine. He set out to analyse the principal animal fluids such as blood, milk, urine, saliva, gastric and pancreatic juices and so on, with the purpose of providing the physician with a basic chemical knowledge of these substances. Although these early analyses made by Berzelius were not ultimate elementary analyses, but only gave the proportions of constituent proximate principles in the animal fluids, they were nevertheless more detailed than any analyses previously made in this field. His results were included in a two-volume textbook of animal chemistry, the first of its kind, published in 1806-8. This book was written in Swedish, Berzelius' native tongue, and although attempts to have it translated into English were unsuccessful, as we shall see, it was still a very important source of inspiration and of information for his contemporaries. For example, Prout drew freely upon it when he gave a series of private lectures on animal chemistry in 1814.

English chemists were able to gain an idea of Berzelius' work from a paper on the analysis of animal fluids which appeared in an English medical journal of 1812. We might do the same, beginning with his treatment of blood. This fluid was drawn from a vein and was examined both immediately and after standing undisturbed for specified periods during which the solids separate from the serum. Berzelius allowed this separation to take place and then proceeded to examine the coagulated part (often called the crassamentum), and the serum separately. The crassamentum was found to consist largely of fibrin, insoluble in cold water but dissolved partially by boiling. When treated with alcohol a fatty substance resembling adipocire was formed. An interesting observation was that this fibrin was capable of reacting with mineral acids in two different proportions. When a small quantity of acid was used a neutral compound soluble in water was formed and with an excess of mineral acid an insoluble acid compound was produced. Nitric acid reacted in a similar way with fibrin— two distinct yellow compounds were formed and at the same time nitrogen was evolved. It seemed that organic substances could combine together according to the laws of definite and multiple proportions in the same way as mineral substances.

Now it was among Berzelius' great ambitions to show that the atomic theory, so successful in the inorganic field, also held good in the sphere of organic and organised substances. The compound 'atoms' involved in organic reactions were clearly much more complex than those found amongst inorganic substances and Berzelius was not at first sure that the same laws could be made to apply to both kinds of matter. In 1812 he said, '... it is very evident that the mode in which combustible bodies combine with one another, ... is totally different from that which prevails among the inorganic productions of nature.' Within two years however, Berzelius had become convinced that the organic 'atoms' must be thought of as made of radicals, each containing several elements in the manner proposed by Lavoisier. The same laws of combination as those already established in inorganic chemistry must then be thought to govern organic compounds.

From his observations in electrolysis Berzelius had been led to suggest that all elements were either electropositive or electronegative and that the charges which the atoms carried were the root cause of chemical affinities as well as the source of heat evolved during reactions. Amongst organic compounds however, the same small group of elements produced at one time a strongly electronegative radical and at other times a neutral compound. It was difficult to see how a single electrochemical theory could be made to account for all the reactions of both mineral and organic substances and yet Berzelius realised that this was at least desirable if not even essential for the future development of the subject. He thought that possibly the complexity of the organic radicals obscured their electrostatic charges and he was also at first ready to accept that the radicals themselves might never be isolated. A great deal more work was needed before the concept of organic radicals could lead to significant developments in the study of structural chemistry.

Returning to Berzelius' study of the blood, we find that he paid particular attention to the colouring matter in the crassamentum. He cut the coagulated blood into thin slices which were dried and triturated with water until no more would dissolve. The solution was deep brown in colour and it coagulated on heating leaving a deposit from which by prolonged boiling with dilute hydrochloric acid, Berzelius was able to extract an iron compound. He found iron oxide in the ash remaining after the colouring matter of the blood had been burned and he suggested that this was the main difference between fibrin and albumen on the one hand and the colouring matter of the blood on the other. The exact manner in which iron was to be found combined in the colouring matter remained a difficult problem which Berzelius was not able to solve. Fourcroy had suggested that the red colouring matter was due to the presence of a solution of the 'sub-phosphate of iron' in albumen, but Berzelius thought that if this were so the iron should be recognisable by the ordinary tests. Since it was not he thought that the iron must be present in some special organic combination. The results of burning the colouring matter in an open crucible seemed to show that this substance also contained carbon, phosphorus, sulphur, calcium and ammonium in addition to iron.

In the serum of human blood Berzelius found albumen and a number of simple salts. The results of this analysis were found to be very similar to those already obtained by Alexander

TABLE III. *Berzelius' analysis of the urine*
Medico-Chirurigical Transactions, Vol. 3 (1812)

Water	933.00	parts by wt.
Urea	30.10	„ „ „
Potassium Sulphate	3.71	„ „ „
Sodium Sulphate	3.16	„ „ „
Sodium Phosphate	2.94	„ „ „
Sodium Chloride	4.45	„ „ „
Ammonium Phosphate	1.65	„ „ „
Ammonium Chloride	1.50	„ „ „
Lactic acid; ammonium lactuate; alcohol-soluble and alcohol-insoluble animal matter	17.14	„ „ „
Earthy phosphates; calcium fluoride ..	1.00	„ „ „
Uric acid	1.00	„ „ „
Mucus of the bladder	0.32	„ „ „
Silex	0.03	„ „ „
	1000.00	

Marcet, physician at Guy's Hospital in London, on the composition of dropsical fluids and this observation seemed to bear out Berzelius' view that all secreted fluids were derived from the blood with only minor changes. Thus, bile, saliva, humours of the eye, mucous and serous fluids, all contained the same salts in the same proportions in aqueous solution as the serum of the blood itself and consequently the only observable changes involved in the formation of these secretions involved merely the relative proportions of albumen. Yet, since the secretions were clearly quite distinct from each other and from the serum of the blood, the changes leading to the formation of each were clearly specific. They were evidently a function of the animal economy and Berzelius thought that no similar changes were to be found anywhere in the inanimate world. He surmised that the nervous system was involved in the process of secretion and that there might well be some form of electricity acting in the secreting organs. Sir Benjamin Brodie and others also investigated the action of the nerves in controlling the vital functions of the body.

In general the secretions were found to be alkaline, but the excretions were acid—a fact which pointed to the truth of the view that the animal economy is chiefly concerned with *oxidation* processes. Excretions also contain a much higher proportion of dissolved salts, as Berzelius' analysis of the urine shows. This was the best analysis of this fluid then available and it remained standard for a number of years—it was still being quoted in 1848. It is interesting to note that Berzelius detected a small quantity of fluoride, but thought that both nitrates and carbonates were absent. In the case of the nitrates it was simply that a more delicate test was needed and this did not become available until much later. Indeed, both qualitative and quantitative organic analysis needed considerable refinement before real advances in animal chemistry were to become feasible.

Chapter 3

Animal Analysis

IT WILL ALREADY have become clear that the animal chemist of the early nineteenth century was in urgent need of an adequate method of elementary organic analysis. The general description of animal substances and their behaviour on boiling or when treated with acids, alkalies or mineral salts in solution may well have been enough to characterise and classify them, but such information is no substitute for a full chemical analysis. Until the time of Lavoisier the only procedure in use for this purpose was destructive distillation which gave rise to a mixture of more or less complex substances called 'proximate principles' and thought to pre-exist in the organic matter. In practice these products depended very much upon the temperature at which the distillation was carried out and in most cases the substances were by no means either simple or unvarying. Nevertheless, the analysis of organic matter was carried out on this basis for a very long time and even when improved methods were introduced, complex 'organised' matter would frequently be first broken down into its proximate principles before these were submitted to ultimate analysis. After this the elementary composition of the organic matter could be calculated.

Reliable analyses depend upon purity, both of the reagents and of the material under analysis. Here the animal chemist was faced with one of his greatest difficulties, for whereas many vegetable and most mineral substances can be purified by precipitation or crystallisation techniques, animal substances are frequently mucilaginous, gummy, albuminous or otherwise impossible to obtain in pure crystalline form. Consequently the animal chemist found it difficult to know when he was working with a single, pure compound. In practice his substances were generally complex mixtures of several compounds. William

Prout made the point when he wrote, '... the utmost care should be taken that the substance operated on be *pure*, a point of greater importance, and frequently of more difficult accomplishment than any other.' Prout went on to say that this single difficulty had caused him more trouble than all the rest put together. It was for this reason that animal chemists sought to break down complex substances such as blood, body tissues, matter of the brain and so on, into their proximate principles before attempting to discover their elementary composition.

Lavoisier's oxidation analyses

After Lavoisier had introduced the definition of the chemical element as the last point in analysis, the idea that the proximate principles of distillation analysis could be regarded as the 'elements' of organic matter was no longer tenable. From his recognition of carbon and hydrogen as the two chief elementary constituents of natural substances it followed that combustion or oxidation analysis must give rise to carbonic acid and water vapour as the principal products, azote remaining unchanged due to its inert nature. The volume of carbon dioxide produced could be found by absorption in caustic potash solution and the water vapour would reveal its own volume by virtue of the fact that it would condense on cooling. The azote would then remain unchanged as a gas. In this way the analysis could be made quantitative and the percentage composition of the organic substance could be calculated in terms of its chemical elements.

Lavoisier's method of combustion analysis was based on these ideas. In his early attempts he used weighed quantities of alcohol, olive oil and paraffin wax which he burned in a bell-jar full of air, confined over mercury. Fresh air or oxygen was added from a second vessel connected to the first by a delivery tube and in this way the combustion was maintained.

Lavoisier found that if pure oxygen was used with volatile substances like alcohol and ether, explosions resulted. The carbon dioxide formed in the combustion process was absorbed in pure caustic potash solution and the water was calculated from the difference between the volumes of the gaseous reactants and their products after combustion. The method was not very accurate—it must have been extremely difficult for example, to measure the volumes of the gases with any degree of precision since the vessels in which they were contained were so wide.

In another method of oxidation analysis which was to be the forerunner of the more accurate later procedures, Lavoisier heated organic substances mixed with mercuric oxide. Some entries in his notebook for 1788 show that he used this method with sugar and that the carbon dioxide evolved in this case was absorbed in *weighed* potash bulbs—a much more accurate procedure. Unfortunately his results were still full of minor errors. For example, he took the ratio of mercury to oxygen to be 92 : 8 parts by weight instead of the more correct value of 100 : 8, and as a result he found only 23 per cent of carbon in his sugar instead of the true value of 42 per cent. In some other experiments of this kind Lavoisier used manganese dioxide or potassium chlorate as oxidising agents instead of mercuric oxide and in this way he laid the foundations of the combustion analysis technique introduced by Gay Lussac and Thenard, developed by Prout and Berzelius and brought to perfection by Liebig and his students.

The analysis of alcohol and ether

Theodore Saussure (1767-1845), the Swiss organic chemist whose main interest was in the study of plant functions, attempted to improve the distillation analysis procedure so as to make it capable of giving quantitative results. He used mixtures of alcohol and water the compositions of which he was able to estimate with reasonable accuracy from specific gravity. One mixture he examined contained 86.2 parts of absolute alcohol to 13.8 parts of water by weight. This liquid he slowly distilled through a red-hot porcelain tube and the products were carefully collected. They consisted of water and a hydrocarbon gas which could be analysed by explosion with oxygen. From this complex procedure de Saussure concluded that alcohol was composed of olefiant gas (ethylene C_2H_4) and water in the proportions of two molecules to one respectively. In fact as with all distillation analyses the composition of the gaseous product depended upon the exact temperature of the porcelain tube and it was necessary to take the mean of a number of tests—the final result was not very reliable.

De Saussure then turned to 'sulphuric ether' (i.e. ordinary ether made by treating alcohol with concentrated sulphuric acid). This he rectified by distillation, first with potash to neutralise any residual acid, then after washing with water he distilled it from fused calcium chloride, accepting only the last third of the distillate for analysis. This sample had a specific gravity of 0.7155 at 68°F. and was therefore nearly pure ether.

'I distilled at a temperature below that at which ether boils, 47 gms of that liquor through a red-hot porcelain tube. This apparatus was similar to that described for the analysis of alcohol. The operation lasted nine hours ...', de Saussure tells us.

In this case he collected several products including a gas, some charcoal and an oil, but there was considerable loss due to the high volatility of ether. The loss was enough to render any experiment of this kind valueless for the determination of percentage composition and de Saussure therefore turned to another method. He decided to employ the technique of exploding ether vapour with oxygen in a procedure similar to that which had been used by Lavoisier.

For this purpose he introduced a small weighed phial containing pure ether into a measured volume of oxygen confined over mercury and detonated the mixture by means of electric sparks. From the change in volume of the mixture after cooling and the volume absorbed when potash was introduced he was able to determine the approximate composition of ether. He found 67.98 parts of carbon, 17.62 parts of oxygen and 14.40 parts of hydrogen by weight, or five molecules of olefiant gas to one of water. De Saussure next went on to apply this explosion technique to the analysis of a number of natural substances including oils, gum arabic, the resin colophony and a few animal oils such as wax from the ears, spermaceti, biliary calculi, pork fat and so on. Some of these substances also contain nitrogen. He gave percentage compositions for the compounds examined and, although his results were not very accurate, his work demonstrated clearly the possibilities with respect to elementary analysis. By applying Dalton's atomic theory these percentages could readily be converted into empirical formulae. Thomas Thomson, as editor of the journal *Annals of Philosophy*, in which de Saussure's early work was published, showed how this might be done, but the formulae so derived left much room for improvement.

Dalton was himself aware of the possibility of analysing volatile organic compounds by exploding their vapours with oxygen. In 1805 he had found that when ten volumes of ether were exploded with sixty volumes of oxygen in a eudiometer, forty volumes of carbon dioxide were produced. From such analyses together with his vapour density measurements, Dalton was able to show that the 'atomic weight' of ether was 20.8 (H=1) and he concluded that ether contained two parts of olefiant gas to one part of water. Alcohol, he thought, contained equal amounts of these two constituents. Later, Dalton examined the gases evolved when whale-oil and spermaceti were distilled. He found a hydrocarbon gas which behaved like olefiant gas on explosion with oxygen but which combined with only half as much chlorine as olefiant gas. In fact, Dalton had discovered butylene (C_4H_8), a new form of olefiant gas in which the *proportions* of hydrogen and carbon are the same as in common olefiant gas but the *quantities* of each element are doubled.

This discovery demonstrates a difficulty which faced all organic chemists, not least the animal chemists themselves. So many natural substances are found to have the same or very similar empirical or molecular formulae that it was clear that very great precision would be necessary in order to distinguish them. Such similarities led to the speculation that it might be possible to pass from one such substance to another by the mere addition or subtraction of very small quantities of matter —a few atoms in fact. The atomic theory was faced with problems in the field of inorganic and mineral chemistry, but the problems were clearly many times greater amongst organic and organised compounds. The occasional presence of nitrogen, sulphur, phosphorus and other elements in animal substances only added to the problems of animal chemists.

Gay Lussac and Thenard

The first successful method of elementary organic analysis was introduced by the French chemists Gay Lussac and Thenard, both of whom had worked with Fourcroy and Vauquelin and were members of Berthollet's private research group known as the Society of Arcueil, founded in 1807. Their method of analysis depended upon the oxidation of organic matter to form carbon dioxide and water, using potassium chlorate as the oxidising agent. The organic substance was first mixed with powdered potassium chlorate, made into pellets and dropped through a special stopcock into a vertical tube heated at its base in a charcoal fire over a spirit lamp. The gases evolved were then collected over mercury and analysed.

In order to be certain that the chemical process was complete Gay Lussac and Thenard tried different mixtures until they found one which left a white residue after heating. In general it was necessary to use a 50 per cent excess of potassium chlorate to be sure of converting *all* the carbon and hydrogen to their oxides. It was also necessary to find the volume of oxygen given off by the potassium chlorate alone and this was done in a separate experiment. When carrying out an analysis several pellets of the mixture of potassium chlorate and the organic substance under test were first decomposed in the apparatus so as to fill it with the same mixture of gases as that to be analysed. After this a number of weighed pellets were introduced into the apparatus and the evolved gases were collected, measured and analysed to find the ratio of carbon dioxide, oxygen, nitrogen and water vapour.

Gay Lussac and Thenard applied this method to a number or organic substances including sugar, gum, starch, milk sugar, albumen, casein, gelatin, oak and beech woods, resin, copal, wax, olive oil, mucic, oxalic, citric and acetic acids and fibrin. In general the results of their analyses were good and were quoted widely. They showed that in sugar, starch and gum the proportions of hydrogen and oxygen were the same as those in water; in resins and oils there was found to be an excess of hydrogen and in acids, an excess of oxygen. Fibrin, albumen, casein and gelatin were all found to contain a similar proportion of nitrogen.

Gay Lussac later introduced an important modification of the procedure in which he used copper oxide as the oxidising agent and fixed the tube in a horizontal position. One source of error in the original method had been that the pellets had to turn as they fell through the greased stopcock and small amounts of the grease adhered to them as they did so. This was then included in the analysis, introducing an error in the percentage of carbon. The use of a horizontal tube eliminated this error but the experiment was still difficult to perform successfully. Since one of the products was water vapour it was necessary to make sure that the apparatus and all the reagents were thoroughly dry, otherwise the proportion of hydrogen found would be seriously affected. In order to prevent this error from occurring Gay Lussac heated the apparatus in a bath of salt solution and at the same time reduced the pressure inside it. He then allowed the air to re-enter the apparatus slowly through drying tubes containing fused calcium chloride. To be certain that the apparatus was quite dry this process had to be repeated twelve to fifteen times and the precautions so taken made the method very tedious and time-consuming. After the experiment the water vapour was weighed, having been absorbed in weighed calcium chloride tubes, the volume of carbon dioxide was found by absorption in caustic potash solution and the remaining gas was counted as azote or nitrogen. Any oxygen in the organic compound was estimated from the difference between the loss in weight of the copper oxide and the total weight of oxygen found in the water vapour and carbon dioxide formed during the process. The method was laborious but reliable and despite its slowness it attracted attention from other organic and animal chemists, some of whom went on to improve its practical usefulness.

Berzelius turns to ultimate analysis

Following his early interests in the proximate analysis of animal fluids which we have already mentioned, Berzelius attempted to achieve a more exact knowledge of the composition of organic and organised matter by applying the techniques of elementary analysis. He began with some relatively simple compounds, which included oxalic, tartaric and citric acids, hoping to be able to give the correct formulae for these compounds by means of Dalton's atomic theory. In order to do this with any degree of reliability however, it was first necessary to improve the experimental techniques of organic analysis and this Berzelius set out to do. By 1814 he had devised a method which was both reasonably convenient and dependable with respect to accuracy.

His method consisted of heating the organic matter with potassium chlorate as an oxidising agent, mixed with some sodium chloride to moderate the reaction, in a hard glass tube closed at one end, encased in an iron tube and supported in a sloping position on a brick, (Fig. I). The water vapour produced during the reaction was condensed in a bulb and the last traces were removed with fused calcium chloride. The remaining dried gases were then collected over mercury in a bell-jar. Carbon dioxide in these gases was then absorbed by introducing a small bulb containing a weighed quantity of potassium hydroxide, the increased weight of which was found. In this way the weights of both water vapour and carbon dioxide formed during the combustion of the organic matter were found and from these weights the proportions of carbon and hydrogen in the organic matter could be estimated. The results from Berzelius' method were reasonably good for these two elements, but the quantities of oxygen and nitrogen in the organic matter were much less easy to determine and further refinements were clearly necessary before more complex animal substances could be analysed.

As we have already indicated, one of the great practical difficulties of this type of analysis was the problem of ensuring that all the reactants and the apparatus were thoroughly *dry*. In the method of Gay Lussac and Thenard for example, the pellets had to be moistened in order to shape them and consequently it was doubly necessary to dry them carefully before proceeding with the analysis. Failure to do so led to enhanced figures for the proportion of hydrogen in the organic matter. Berzelius overcame this and was able to achieve a high degree

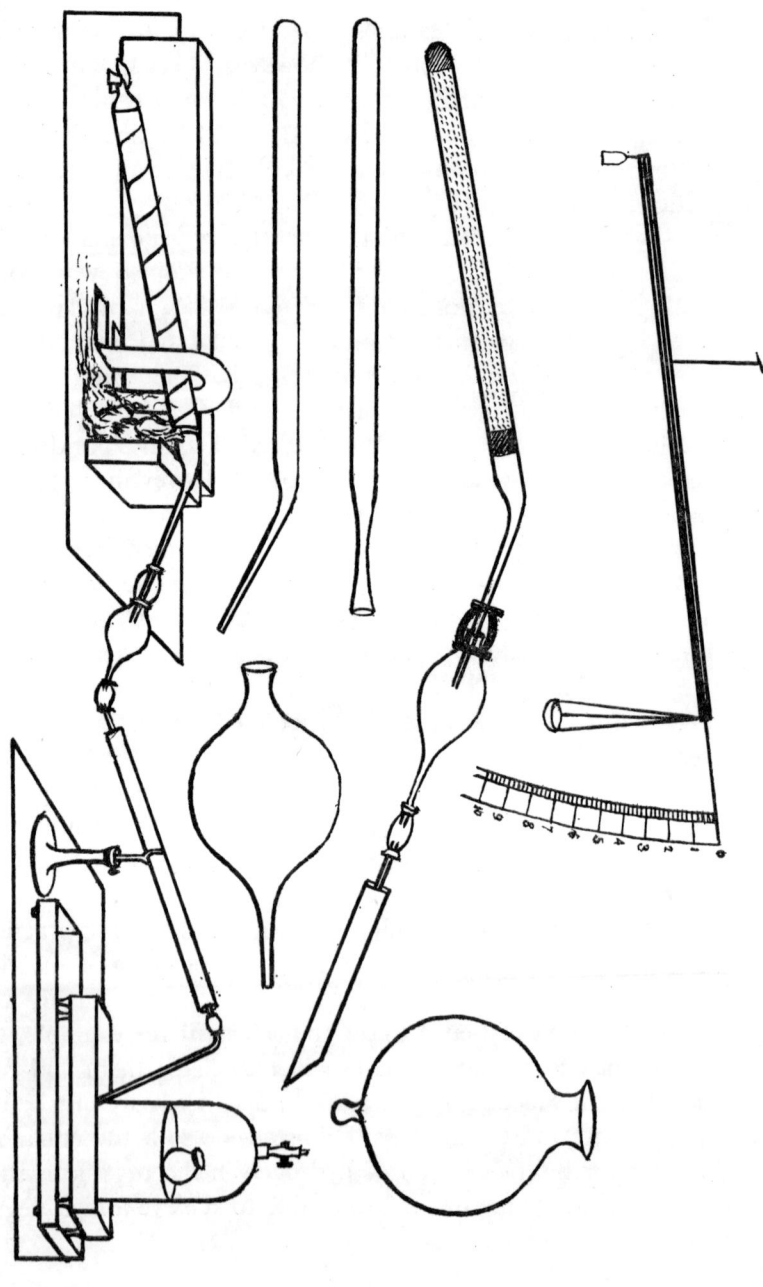

Figure 1. Berzelius' apparatus for elementary organic analysis. cf. Plate 3. (From *Annals of Philosophy,* 1814.)

of dessication by pulverising his reactants in a hot mortar. Whilst doing this he took good care to wear gloves and to keep his breath away from the mortar since he had found that even the moisture from the hands and the breath was enough to upset the results.

Berzelius also tried to control the rate of the reaction by means of a screen which he moved slowly along the tube limiting the effects of the fire to one part of the tube at a time. Inside the tube he packed a good excess of pure potassium chlorate at each end so as to ensure a good excess of oxygen. This would serve both to oxidise the organic matter completely and to carry the gaseous products of combustion over into the bell-jar, but it also had the effect of masking the quantities of oxygen and nitrogen present in the organic matter.

Berzelius had also tried to analyse the lead salts of oxalic, and tartaric acids using lead dioxide as the oxidising agent. The reactions in these cases turned out to be too rapid and large volumes of gases were evolved which made it difficult to keep the apparatus gas-tight. It was to overcome this problem that Berzelius had introduced his methods of mixing the oxidising agent with sodium chloride to moderate the reaction and of heating only part of the tube at a time. He found a further source of error when using lead dioxide however, arising from the fact that this compound itself always gave off small quantities of carbon dioxide and water vapour when heated. These added to the proportions of carbon and hydrogen found in the organic matter and their presence had to be allowed for in calculating the results. Berzelius tried to eliminate this source of error as far as possible by insisting upon the use of the purest reagents and scrupulous care in drying them and in carrying out the procedure so as to make it possible to repeat the determinations exactly. Analysis is of course of no value less the results are repeatable within a narrow margin of error. Berzelius' work represents an important advance in elementary organic analysis and in the awareness of its problems.

William Prout and the analysis of organised products

The two distinct aspects of Berzelius' work in animal analysis are to be found also in the work of other nineteenth century animal chemists. William Prout, perhaps the most important amongst contemporary English animal chemists, wrote in 1820,

> I have for several years been engaged in the analysis of organised products, and have at length extended my researches to almost every distinct and well-defined substance. The results, when compared with one another, are most interesting, and seem to throw no small light not only on the nature of chemical compounds in general, but upon many important points connected with animal and vegetable physiology and pathology.

As a physician Prout was interested in the applications of chemistry to physiology and he believed that chemical analysis of organic and organised matter could be of immense value to the physiologist. He took up Gay Lussac's later suggestion that copper oxide might be used as an oxidising agent in combustion analysis—although it was known that this substance is hygroscopic and tends to absorb gases at high temperatures. Prout designed several modifications of an apparatus similar to that of Berzelius but using copper oxide in place of potassium chlorate. In some forms of the apparatus the combustion tube was kept horizontal whilst in others it was held in a vertical position (Fig. II). He applied his method to the analysis of urea, urea nitrate, sugars and lithic (i.e. uric) acid, later extending his researches to albumen, cystic oxide (cystine), and oxalic acid. Like Berzelius, Prout was also interested in the atomic theory and he tried to show how these substances could be related to each other through their chemical formulae. On the whole his analytical results were good and this enabled him to propose reasonably accurate formulae, although some of his attempts

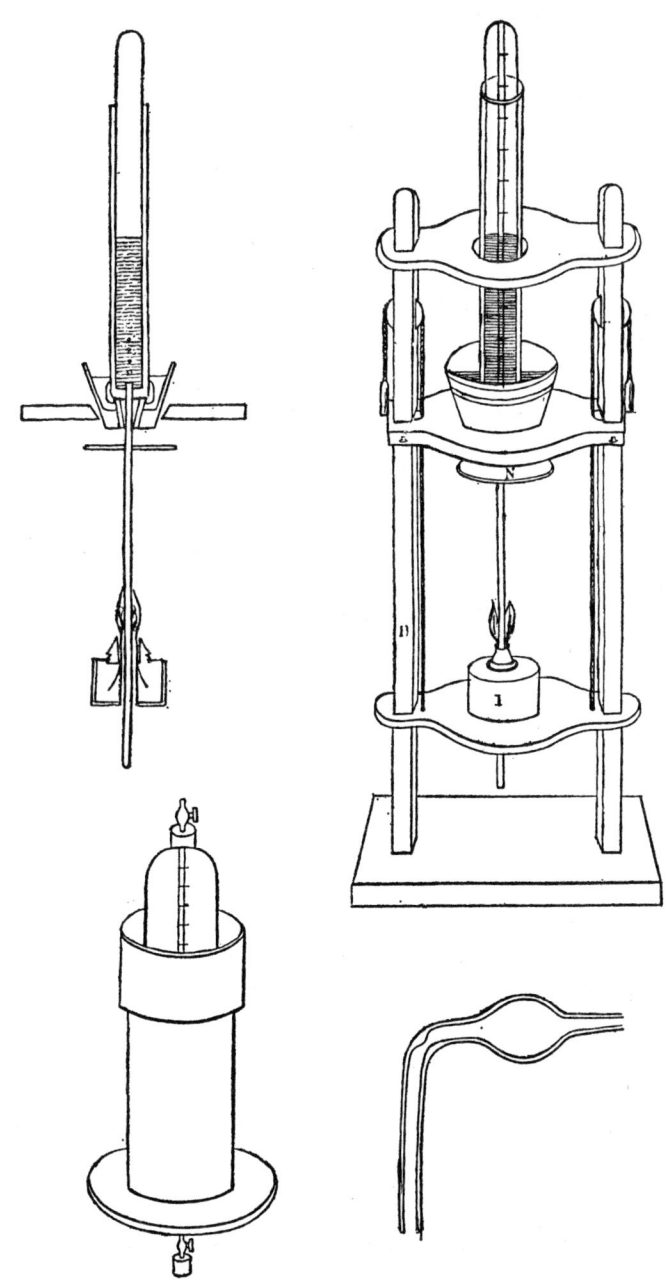

Figure II. Prout's apparatus for the analysis of Organised Substances.
(From *Annals of Philosophy*, 1820.)

to relate these to each other now seem to be very naïve. For example, he thought that he could show two 'atoms' of sugar to be equivalent to one 'atom' of urea, whilst lithic acid, was equivalent to three 'atoms' of sugar, cystic oxide to four and albumen to six. This would then help him to explain how organised substances in the body were derived from and related to an important item of the diet. Prout was attempting to simplify the complexities of nutrition to an unjustified extent and the fact that he made no reference to the nitrogen present in urea, lithic acid and albumen, nor to the sulphur in cystic oxide indicates some of the inadequacies of organic analysis at this time.

Prout laid great emphasis on purity both in his reagents and in the organic substances under examination. He was one of the first to obtain a really good sample of crystalline urea from the urine and his method of doing so illustrates the trouble he was prepared to take in order to attain the highest possible degree of chemical purity. Fresh urine was first evaporated to the consistency of a syrup and after cooling pure concentrated nitric acid was added until the whole had turned to a dark crystalline mass. These crystals were washed with cold distilled water and a solution of sodium or potassium carbonate was added to neutralise the nitric acid. The solution so formed was evaporated until it would crystallise to form sodium or potassium nitrate crystals and a solution containing impure urea. The crystals were filtered off and the solution was then boiled with animal charcoal in order to decolourise it. The mixture was again filtered, evaporated to half bulk and allowed to crystallise. These crystals, which were impure urea, were then filtered off and dissolved in boiling alcohol. After filtering the hot solution so formed the urea was allowed to crystallise from it in four sided, transparent crystals of fair purity.

In 1827 Prout introduced a modified method of organic analysis in which oxygen gas was passed repeatedly over a

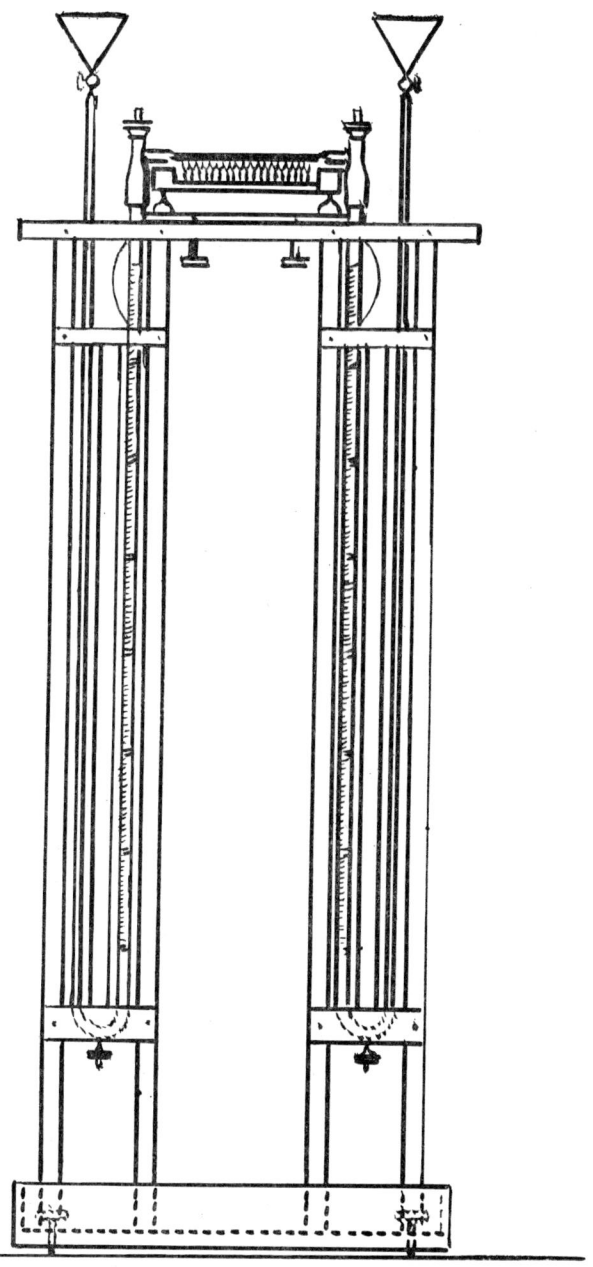

Figure III. Prout's later apparatus for organic analysis using oxygen. (From *Philosophical Transactions Roy. Soc.*, 1827.)

heated mixture of the organic substance and copper oxide, (Fig. III). By carefully regulating the flow of oxygen the combustion of the organic substances could be made complete, but the results of this process were not in fact more accurate since only the volumes of residual oxygen and carbon dioxide were measured and the hydrogen was estimated by difference. Prout applied his new method to organic substances containing carbon, hydrogen and oxygen. He argued that since one volume of carbon dioxide is formed from one volume of oxygen, there would be no volume change when the carbon of the compound was oxidised. However, if the proportions of hydrogen to oxygen in the compound were greater than 2:1 the volume of residual oxygen would be decreased, it would be increased if the proportions were less than 2:1 and in those compounds in which the proportions of hydrogen to oxygen were exactly the same as in water, as in carbohydrates— Prout's 'saccharine substances'—the volume of oxygen would remain unchanged. From such considerations Prout proposed to determine the formulae of the substances which he analysed merely by making careful measurements of the carbon dioxide formed and the residual oxygen. He used the method principally for the analysis of the main components of the food.

Liebig and the Giessen School

Each of the methods of organic analysis so far mentioned had its merits, but each was especially suited to the skills of the particular chemist who had devised it. What was really needed was a general method which could be adapted for use in any branch of organic chemistry and which could be carried out with ease and accuracy by any competent chemist. Such a method would have to be simple, speedy and reliable; it should be so devised that it could be made a matter of routine in any chemical laboratory. It is to Justus von Liebig that we owe the method of organic analysis which had these features and which, in his laboratory at Giessen was developed to such a degree of perfection that it could be applied to several hundred compounds each year.

Liebig, who became professor of chemistry at Giessen at the age of twenty-one, had worked with Gay Lussac in Paris and had learned from him the method of combustion analysis which we have already described. Indeed, Liebig's work was based on what he knew of the earlier work of Gay Lussac and Berzelius. Liebig used copper oxide as his oxidising agent and prepared it himself by heating copper carbonate or nitrate. The latter gave a harder and less hygroscopic product and the same sample of copper oxide could be used repeatedly by moistening it with nitric acid and re-igniting. Liebig's combustion tubes were made of lead-free Bohemian glass which was capable of withstanding the temperature of the charcoal furnace without cracking or fusing.

When carrying out an analysis the procedure was as follows. The organic material was first carefully weighed and then thoroughly mixed with pure copper oxide by grinding to a fine, uniform powder in a hot, porcelain mortar. One end of the combustion tube was sealed and a short length of it was filled with pure copper oxide. The main part of the mixture was then introduced, the mortar and weighing bottle were 'rinsed' with pure copper oxide and the rinsings added to the tube. Finally a short length of the combustion tube was again filled with pure copper oxide and the whole supported in a sloping position in a furnace, so that the water formed during the reaction would run down to the cooler end of the tube. Oxygen and carbon dioxide evolved in the reaction carried the water as vapour over into weighed drying tubes filled with fused calcium chloride. Liebig, like Berzelius, regulated the rate of reaction by moving an iron sheet slowly along the tube so that only a short part of it was being heated at a time, (Fig. IV).

The weight of carbon in the organic substance could be

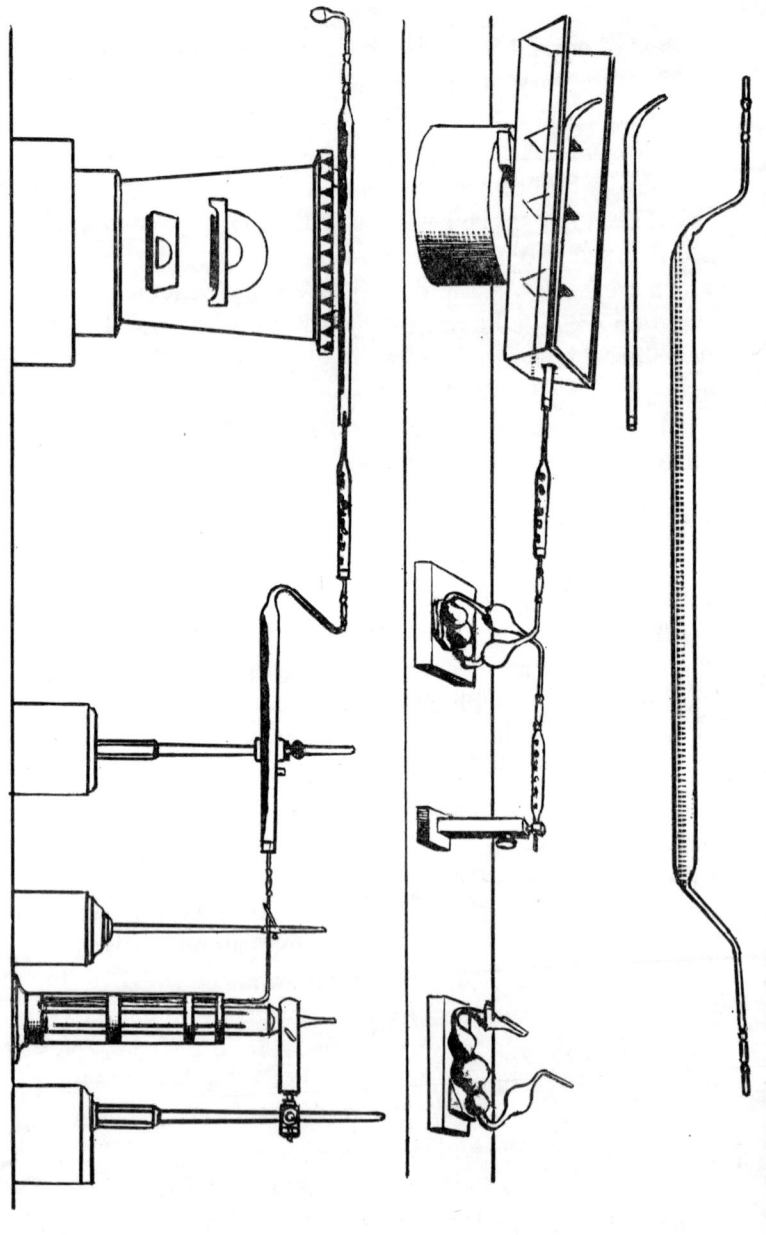

Figure IV. Liebig's apparatus for elementary organic analysis. cf. Plate 4. (From *Annales de Chimie*, 1831.)

found from the weight of carbon dioxide produced during the combustion and absorbed in weighed potash bulbs. Similarly the weight of hydrogen could be found from the increase in weight of the drying tubes filled with fused calcium chloride. The main sources of error were the possibility of incomplete combustion and the effects of the atmosphere in bringing added carbon dioxide into the apparatus or in evaporating some of the water from the potash solution.

After the determination of carbon and hydrogen, nitrogen had to be estimated in a separate experiment. A small quantity of the organic substance was heated in the combustion tube with a very large excess of copper oxide so that a volume of gas consisting of carbon dioxide and nitrogen was produced and collected. The proportions of these two gases in this mixture was then determined by absorbing the carbon dioxide in potash and from a knowledge of the exact quantities of carbon dioxide and water vapour given off from a known weight of the organic substance, the weight of nitrogen present could be calculated.

Liebig's method also took account of those cases in which the organic substance might contain halogens or sulphur. The halogens were to be estimated by heating the organic matter with quicklime, dissolving the calcium halides so formed in pure distilled water and then precipitating them with silver nitrate. After this the silver halide could be filtered off, washed, dried and weighed in the normal manner. Sulphur was oxidised to sulphates by fusion of the organic matter with potassium hydroxide mixed with a small proportion of potassium nitrate to act as an oxidising agent. The sulphate was then precipitated as barium sulphate; the precipitate was washed, ignited and weighed.

All the procedures described by Liebig were relatively simple, straightforward experimental methods which were well within the capabilities of the average analytical chemist. This made the processes of elementary organic analysis available to those whose skill in practical chemistry was only moderate. At Giessen Liebig encouraged his students to carry out such analyses in large numbers and between 1825 and 1852 his laboratory became famous throughout the world for the work in organic chemistry carried out there. With Liebig's work the point had been reached when animal chemistry could be supplied with reliable analytical data and by means of the atomic theory it became possible to calculate the empirical formulae of animal substances. Thus the outlines of the subject began to take shape in terms which could be understood by all who had a basic knowledge of chemistry.

Nevertheless, one major problem remained concerning the chemistry and physics of life. Were animal functions wholly chemical or were they in fact governed by some undefined life-force? Animal substances and functions were so complex that it was difficult to maintain that they could be explained without the aid of vital force and the increasing knowledge of chemical composition derived from animal analysis did little to clarify this question. Even Liebig himself was a vitalist, though he thought that the vital force would be found to obey laws similar to those governing the forces of gravity, electrostatics and magnetism. Vitalism in one form or another continued to exert its influence on animal chemistry until at least the end of the nineteenth century and even today the idea is not entirely non-existent, although it does not now find acceptance amongst scientists.

Chapter 4

Chemists, Physiologists and Doctors

To the physician of the nineteenth century, animal chemistry seemed to offer invaluable aid in the diagnosis and treatment of disease. Any physician who could interpret the healthy functions of the organs in terms of physics and chemistry, would clearly be in an excellent position to decide what was wrong when disease occurred. His medicine would then be based upon rational deductions made from scientific observations and the results of chemical research; his diagnosis would be more accurate and his remedies more likely to have the desired effect. With these thoughts in mind many physicians of the nineteenth century turned to animal chemistry in the hope that they might be able to solve some of their more difficult problems. For example, diseases such as gout, diabetes, stomach disorders and urinary calculus were thought to arise mainly or even solely from chemical causes and animal chemistry seemed to offer a real prospect of success in the treatment of these conditions.

A medical problem for animal chemists

Most eighteenth-century chemists and physiologists were unable to conceive a state of life in the absence of the vital force and there was always some doubt as to whether life functions could really be explained purely in chemical terms. Hence, in view of these reservations it is not surprising to find that some of the early progress in animal chemistry as applied to medicine, was made in connection with the study of mineral deposits in the body—a branch of the subject which seemed nearest to the more familiar inorganic or mineral chemistry. In 1797, W. H. Wollaston wrote a chemical account of some of these deposits in which he said,

> If in any case a chemical knowledge of the effects of diseases will assist us in the cure of them, in none does it seem more likely to be of service than in the removal of the several concretions that are formed in various parts of the body.

Urinary calculus or bladder stone, had been a scourge throughout the eighteenth century and many attempts had been made to find an effective remedy for this painful condition. Many patients were at length forced to submit to the surgical operation of lithotomy, which was both difficult and dangerous. Before the introduction of anaesthetics in the mid-nineteenth century, the operation was a daunting prospect for even the most intrepid patient and other means of relief were therefore constantly being sought. The most promising alternative seemed to lie in the strong alkalies which were often given as medicines, a famous example of which is to be found in the remedies of Mrs Joanna Stephens for which the sum of £5000 was paid out of public funds by order of Parliament in 1739. When Mrs Stephens' remedies were published it appeared that they contained a high proportion of lime. They were made from calcined egg-shells and snails, mixed with a variety of dried vegetable products and finely powdered. The medicine was taken either as a powder or mixed with water and there was also a method of making 'pills' from it by mixing the powder with soap and honey. Clinical tests to assess the real value of these mixtures were at once made by a group of eminent medical practitioners, which included William Cheselden the famous lithotomist, whilst Stephen Hales tried to identify the active chemical ingredients.

Robert Whytt, professor of medicine at Edinburgh in the eighteenth century, had advocated the direct injection of lime-water or soap solution into the bladder, but this process some-

Plate 3. Berzelius' apparatus for elementary organic analysis.
(*Reconstruction*)

Plate 4.   Liebig's apparatus for elementary organic analysis.

(Reconstruction)

times had very undesirable effects. It was in order to discover the relative merits for medicinal purposes of the so-called 'mild' alkalies, that Joseph Black began his classic series of researches on magnesia and lime, published in 1756. This work resulted in the discovery of 'fixed air' or carbon dioxide and led to a clearer understanding of the relations between the alkaline earths and their carbonates. It also showed that gases could form a definite part of some solid substances and could be transferred from one such compound to another.

Thus the search for a cure for bladder stone had already been the subject of a large amount of chemical work and had led to discoveries of some importance when, in 1766, Scheele announced the discovery in urinary stones of a 'concrete acid hitherto unknown'. Scheele found that this substance (really uric acid) was only sparingly soluble in water, showed no reaction with dilute mineral acids, but was readily soluble in alkalies. It gave a characteristic pink or purple colour with concentrated nitric acid and since it was found to be present in most, if not all bladder stones, Scheele's acid principle came to be called 'lithic acid', (Greek *lithos*=a stone).

The results of this work did not become widely known for a decade or more, but from about 1786 onwards Fourcroy and Vauquelin made numerous experiments on calculi of all kinds. They found that Scheele's acid principle was only one amongst a variety of components which included the ammonium salt of this acid together with calcium oxalate, calcium phosphate and magnesium ammonium phosphate. By 1800 the French chemists had examined a total of about 600 stones most of which were seen to have been formed by the deposition of salts in layers upon a nucleus of Scheele's acid or calcium oxalate.

About the same time George Pearson, who had been a pupil of Joseph Black in Edinburgh and who later became senior physician at St George's Hospital, London, treated urinary calculi with sodium hydroxide and so obtained from them a compound which was present in all but six out of two hundred stones examined. This was really Scheele's acid but Pearson thought that it was an 'animal oxide' for which he proposed the name 'ouric oxide'. Fourcroy however, was able to show that Pearson's oxide reacted as an acid upon the addition of water and he suggested that the name should be changed to uric acid—a proposal which was generally adopted.

Another physician who applied his chemical techniques to the analysis of calculi was W. H. Wollaston. Using very small particles of such stones Wollaston tested them by means of their reactions in the flame of the blowpipe and their behaviour in the presence of drops of acids and alkalies. His main object was to develop a method by which the medical practitioner could apply chemical tests in making his diagnoses; chemistry was to be applied so as to improve the precision of medical practice. Wollaston found, as Fourcroy and Vauquelin had done, that in addition to uric acid, calculi also contained calcium salts such as the oxalate and phosphate, as well as magnesium ammonium phosphate. In 1810, Wollaston announced the discovery of a new constituent in bladder stones which he named 'cystic oxide'. This was later renamed cystine by Berzelius.

Whilst these studies of the chemical nature of calculi were proceeding, interest in the possibility of using solvent medicines, either to prevent the formation of calculi or to dissolve those already formed, also continued. Attempts to relieve the suffering of those afflicted with 'the stone' had for long been founded upon the hope of discovering an effective solvent and in 1808 Everard Home returned to this when he suggested the use of mild alkalies to dissolve the uric acid before it could form a deposit in the kidney. At the suggestion of Charles Hatchett, W. T. Brande, later to become professor of chemistry at the Royal Institution, set out to test the effects of magnesia in this connection. It is interesting to note that Brande carried out his tests on patients suffering from the stone and he found that although magnesia reduced the quantity of uric acid deposited, it caused an increased deposit of phosphates, particularly magnesium ammonium phosphate. The latter could be controlled by means of citric, tartaric or even carbonic acid, but excessive use of these acids led to the reappearance of the uric acid deposit. Clearly the chemistry of urinary deposits required careful handling and none of these solvent medicines could be held to be fully effective in preventing precipitation from the urine of one or another of its solids, whenever these were in excess. Brande also analysed 150 stones in the Hunterian Museum of the Royal College of Surgeons in London. He found nearly all the usual constituents in these stones but concluded from his tests that ammonium urate could not occur; this was an error which was to be corrected later, in 1819, by William Prout.

The Animal Chemistry Society

Home, Hatchett and Brande were all members of the Society for the Improvement of Animal Chemistry, set up in 1808 by a small group of Fellows of the Royal Society. These men included physicians, chemists and surgeons and they met as a dining club in the homes of Hatchett and Home, alternately, in order to read papers and discuss topics of mutual interest. Unlike some other specialist groups (e.g. the Geological Society, founded in 1807), the Society for the Improvement of Animal Chemistry remained firmly attached to the Royal Society itself. The work done by its members was submitted to the parent body and at least sixteen papers by members of the Animal Chemistry Society were published in the *Philosophical Transactions* between 1809 and 1825. As a result of their work on animal chemistry, Benjamin Brodie and William Brande were both awarded the Copley Medal of the Royal Society, Brodie in 1811 and Brande two years later. Nevertheless, it is unfortunately true to say that this Society did not bring about any lasting or outstanding improvements in the subject. The difficult problems of effective collaboration between chemists, physiologists and anatomists remained.

Berzelius, like many other nineteenth century chemists, had been trained as a physician and he realised the great value which animal chemistry could have for medicine. He therefore set out to write a comprehensive textbook on the subject, only to discover that the available information was both scanty and unreliable. This led him to make his own analyses of animal substances which he then published between 1806-8 in a two-volume work entitled *Foreläsningar i Djurkemien*, or *Lessons in Animal Chemistry*. We have already sampled the kind of discussion which this book contained when we considered Berzelius' treatment of animal fluids such as blood and urine in Chapter II. In 1808 when the second volume of this work appeared, Berzelius sent a copy of it to Humphry Davy, a founder member of the Animal Chemistry Society and a pioneer of agricultural chemistry. Although the book was in Swedish and Davy was unable to understand it, he nevertheless realised its importance and suggested that it would be a worthy objective for the Animal Chemistry Society to publish an English translation. We are told that at one of the early meetings of the Society, four of the members subscribed a total of £80 towards the cost of this project. William Brande was asked to check the translation which was not made by a scientist, but Berzelius, who saw parts of it during his visit to London in 1812, thought it very poor and said that it was 'fully unreadable'.

Alexander Marcet

Brande once claimed that the Society for the Improvement of Animal Chemistry were the first to bring Berzelius into notice in Britain. Such a claim really belongs with more justice to Alexander Marcet, a physician at Guy's Hospital in London. Marcet was an animal chemist, though not a member of the Animal Chemistry Society, who became acquainted with Berzelius in 1812 when the two chemists worked together on the preparation and properties of carbon disulphide. A mutual respect developed between them which grew into a lifelong friendship maintained through frequent correspondence. Marcet was of Swiss origin but had settled in England and had taken British nationality. He was personally acquainted with many European chemists and it was for this reason that he was elected Foreign Secretary of the Medico-Chirurgical Society, founded in 1805. Marcet tried repeatedly without success to find a publisher for an English translation of Berzelius' book on animal chemistry. It seems that such specialised works were very hard to sell at that time and publishers could not be persuaded to take the financial risks involved. Although he was unable to find a publisher for the whole work, Marcet was able to persuade Berzelius to submit the paper on animal fluids analysis which was published in *Medico-Chirurgical Transactions* in 1812.

When he retired from the Presidency of the Swedish Academy of Science in 1810, Berzelius delivered a long lecture on the history of animal chemistry. This was translated into English by Gustavus Brunmark, chaplain to the Swedish Legation at the Court of St James's and it appeared in 1813 under the title, *A View of the Progress and Present State of Animal Chemistry*. In this case the translation was checked by the physicist and physician Thomas Young and the Quaker pharmacist William Allen—Young included a detailed extract from it in a review of current medical literature, published in 1813. This essay served the useful purpose of recapitulating the history of animal chemistry to 1810, but Berzelius also hoped that it might pave the way for the proposed English translation of his larger treatise, which as we have seen never appeared.

His animal chemistry was later included in his great Treatise of Chemistry which was translated into both German and French (though not into English) and so became well-known throughout the whole of Europe.

By his studies of animal substances, coupled with improved methods of organic analysis to the results of which he applied Dalton's atomic theory, Berzelius did more than any other early nineteenth-century chemist to advance animal chemistry. His great authority as a chemist lent support to his work and he came to be regarded as the leading animal chemist of his time—a position which he retained unchallenged until the advent of Liebig.

Marcet, who like George Pearson, had studied under Joseph Black at Edinburgh, was also interested in the application of chemistry to medicine. Part of his duties as a physician at Guy's Hospital included the instruction of medical students and he acted as a lecturer in chemistry at the medical school there between 1807 and 1820. It was usual at Guy's to illustrate the chemical lectures by means of demonstrations and experiments. This was not then a common practice; its novelty so impressed Berzelius that he introduced it into his own teaching.

Among the students at Guy's there was a keen interest in the causes and chemical nature of urinary calculi because this was a disease which they could all expect to encounter once they embarked on medical practice. Although, as we have seen, there was by this time some reliable information on the subject, little attempt had been made to organise it into a treatise and this Marcet determined to do. His book, which was entitled *An Essay on the Chemical History and Medical Treatment of Calculous Disorders*, was published in 1817. It was the first attempt to deal systematically with all the available information on human calculi in a form suitable for the use of medical students and physicians. The book included descriptions of nine types of calculi together with simple chemical tests by means of which they could be positively identified. Marcet

Summary of Marcet's Tests for Calculi

	Name of Calculus	Reaction with Blowpipe	Reactions with: Acids	Alkalies	Appearance	Chemical Composition
1.	Lithic acid (uric acid)	Blackens; burns with charac. smell leaves a little white alk. ash	Insol. gives pink/purple colour with conc. HNO_3	Soluble	Smooth surface usually in layers inside	Uric acid
2.	Bone Earth	Infusible in the blowpipe	Sol. when powdered. Calc. Ox. is pptd on adding Ammon. Ox.	Insol.	White; friable	Calcium Phosphate
3.	'Triple phosphate'	Ammonia and a white res. on heating	Insol.	Sol. ammonia evolved	White; crystalline	Mag. Ammon. Phosphate
4.	Fusible	Melts to a pearly, white globule	Sol. in HCl Calc. Ox. pptd on adding ammon. Ox. Filter. Add Ammon. Hydrox. to ppt. magnesia	Partially sol. ammonia evolved	White	Mixture of 2. and 3.
5.	'Mulberry'	Swells up on heating. Calc. oxide left. White, strong alk.	slowly dissolves	Insol.	Rough brown like a mulberry	Calcium oxalate (mainly)
6.	Cystic oxide	Decomposes charac. smell.	readily sol.	Readily sol.	Waxy	Cystic ox. (cystine)
7.	mixed					
8.	layered					
9.	Xanthic oxide	Burns charac. smell	—	—	Yellow	Xanthic ox.

based his tests upon the techniques introduced by Wollaston. They required only the most minute quantities of materials and involved methods remarkably similar to those of modern semi-micro chemical analysis. A portable set of small-scale apparatus, including blowpipes, dropping bottles and a small spirit lamp was described by Marcet. It was intended to make possible the performance of these tests at the bedside of the patient. The *Essay* also contained one important chemical discovery—that of 'xanthic oxide', now called xanthine, a compound still found as an occasional constituent of bladder stones. All this information together with descriptions of the symptoms and suggested treatment was set out in an ordered manner and the result was a useful textbook of pathological chemistry.

At Guy's Hospital in the early nineteenth century there was plenty of encouragement for young physicians who were interested in animal chemistry. The famous surgeon Astley Paton Cooper was well aware of the importance of the subject and often asked young physicians to undertake specific pieces of chemical research. He would then use their results, quoting them in his lectures or in his books. For example, Alexander Marcet, at Cooper's request, made chemical analyses of chyle from two dogs one of which had been fed on meat and the other entirely on vegetable matter. William Prout also investigated both chyle and chyme in the same connection and the results of all these tests were used by Cooper in his lectures at the Royal College of Surgeons in London.

Respiration and nutrition

Prout, who is best known for his suggestion that the atoms of all heavier elements are composed of hydrogen (Prout's Hypothesis), made important contributions to animal chemistry as we have seen. Some of his earliest published experiments concerned the process of respiration. He set out to determine the variations in carbon dioxide output of a given individual—himself in these experiments!—during the twenty-four hour cycle. For a period of three weeks in August 1813, Prout stuck carefully to a strict regimen of diet, exercise and sleep, making measurements of his respiration rate and products once every hour during the day and at intervals throughout the night as well. He extended these investigations to include the effects of eating, drinking alcohol and tea, emotional states and the presence or absence of the sun. The largest quantities of carbon dioxide appeared to be evolved between 11 a.m. and 1 p.m., whilst the smallest were produced during the night between 8.30 p.m. and about 3.30 a.m.

Already in 1808, Allen and Pepys had made experiments to measure the volumes of inspired and expired air. In a series of nine experiments it was found that an average of 20-30 cc. of air were absorbed from a volume of three to three and a half litres and the expired air contained 8-10 per cent of carbon dioxide measured by absorption in lime water. In a second type of experiment, the subject was made to breathe the same volume of air many times over. The proportion of carbon dioxide still did not rise above 9.5 per cent, but a further 6 per cent of the original oxygen was found to have been replaced by nitrogen. From these results it seemed that whilst part of the oxygen was converted into carbon dioxide volume for volume during respiration, another smaller part was absorbed by the blood and replaced by nitrogen in the expired air.

The fate of nitrogen in respiration was an interesting problem which Allen and Pepys went on to examine. They tested the effects on a guinea-pig of breathing a mixture of 78 per cent hydrogen and 22 per cent oxygen for a long period. The animal gave off some nitrogen and it was found that some of the oxygen absorbed was not replaced by carbon dioxide. A smaller porportion of carbon dioxide was produced when the animal was asleep than when awake. The results obtained by Allen and Pepys were generally taken to be an accurate account of the chemical changes produced on the air by the process of respiration and they were frequently quoted in the nineteenth century as a model of experimental technique in physiological chemistry.

As we have already seen, one of the difficulties facing the animal chemist was the lack of a reliable and convenient method of organic analysis. Prout expended much effort in attempts to solve this problem, beginning with an analysis of the excrement of the boa constrictor. This material was found to be about 80 per cent uric acid and Prout wondered whether the snake which had produced this matter had become ill whilst in captivity. Wollaston had earlier shown that *guano*, bird droppings from South America, also consisted largely of uric acid and Prout's results were confirmed about 1818 by similar analyses made by John and Edmund Davy. Between 1815 and 1820, Prout tried a number of different analytical methods until he was able to claim that he had analysed almost every well-defined substance and that his researches threw light not only upon the nature of chemical compounds in general but also upon many important points of animal and plant physiology and pathology. Prout thought that the apparently simple numerical relationships which he could trace between some of the components of the food and excretory products found in the urine might well lead to explanations of the ways in which disease occurred in the body and ultimately to methods of prevention or cure. He saw that there must be direct links between all the vital processes.

Prout's interest in animal chemistry had begun in 1814 when he gave a series of lectures on the subject in his own house. He emphasised particularly the importance of urine analysis and the connection between the urine and digestion. He was the first to prepare a chemically pure specimen of urea and make analyses of a variety of sugars. Later, he gave a table of 'molecular weights' which seemed to indicate a simple numerical relationship between sugar and some of the main components of the urine.

TABLE V	mol. wt.	sugar=1
sugar	18.75	1
urea	37.50	2
uric acid	56.25	3
cystic oxide	75.00	4
oxalic acid	112.5	6

Prout was always very keen for physicians to take up animal chemistry because he thought that it was so valuable as a means of improving diagnosis and the understanding of the causes of disease.

With regard to digestion, he thought that it occurred in distinct stages, each of which could be allocated to a specific part of the alimentary canal. Thus, digestion occurred in the stomach, followed by conversion into chyme in the duodenum, into chyle in the lacteals, and finally into blood in the lungs and arteries. It seemed to Prout that the chemical properties of chyme, chyle and blood formed a continuous series and the main problem was to indicate at what precise stage the food material became *vitalised* so as to form part of the living tissues. Prout thought that it might possibly be brought about by bile and pancreatic juice, although since the conversion processes were completed in the lungs where the chyle was finally oxidised into blood, this too must clearly play an important part in vitalisation. In any case the processes of digestion and respiration were seen to be linked through the medium of the blood. During respiration, oxygen of the air was known to be converted into carbon dioxide, volume for volume. Excess carbon in the food was thus used up in a process similar to combustion. All this had been worked out by Prout before 1820 and his ideas about the digestive process and its place in the animal economy seemed to be so well defined that he was prepared to collect them together for publication in a book the impending appearance of which was twice announced in 1823.

Hydrochloric acid in gastric juices

Before this book was completed however, Prout made a very important discovery which made his earlier ideas obsolete and caused him to change his mind. He therefore gave up the idea of publishing his views until he had reconsidered the problem in the light of his startling discovery that the principal acid in gastric fluid was in fact hydrochloric acid. Prout confirmed this for gastric fluids in the rabbit, hare, horse, calf and dog, together with some cases of dyspepsia in humans. The announcement, made in 1824, that such a strong mineral acid existed in the stomach, instead of the much milder lactic or acetic acids which had earlier been thought to cause the acidity of gastric fluid, caused a sensation amongst chemists and physiologists. Many received the suggestion with scepticism and sought the 'safety' of traditional assumptions which seemed to be more reasonable. However, Prout's observations were soon to be corroborated by other workers. In Heidelberg, Friedrich Tiedmann, professor of anatomy and physiology, and Leopold Gmelin, professor of medicine and chemistry, collaborated in writing a book on digestion which was published in 1826. This book, which treated digestion both from the chemical and from the physiological point of view, contained confirmation of the presence of hydrochloric acid in the gastric juice during digestion, although it was thought that other acids (e.g. acetic acid) were also present.

Tiedmann and Gmelin examined the chemical composition and properties of the various animal fluids taking part in the digestive process, including saliva, pancreatic juice, bile and gastric fluid as they were found in all the main types of animals. In order to determine what happened to foods as they underwent the changes brought about by digestion, Tiedmann and Gmelin fed dogs, cats, horses, ruminants and so on with single foods, hoping by this means to trace the fate of each kind of food in turn. Some results of value followed, including the observation that fibrin and albumen passed virtually unchanged into the chyle, that there was a definite limit to the quantities of fats (e.g. butter) which the blood could absorb and that starch was progressively converted into sugar as digestion proceeded. The main purpose of the whole process however, still seemed to be that of 'animalisation'. For all their careful chemical analyses, Tiedmann and Gmelin still found it necessary to postulate some vitalising agency in the living stomach as an integral part of the process of digestion.

It had been shown that the influence of the nerves was needed for the vital functions of the stomach to operate. When the nervous force was removed by cutting the appropriate nerves, the process of digestion ceased, but it could be restored again by means of electric shocks. From this it seemed that vital forces in the stomach might be electrical in nature and might depend upon the functions of the nerves. Nineteenth century physiologists investigated this aspect of vitalism, as we shall see.

Tiedmann and Gmelin had worked together on the study of digestion in an attempt to secure a prize which had been offered in 1823 by the Paris Academy of Science for the best work on the chemistry of the digestive processes, and at the same time a rival attempt by Leuret and Lassaigne, two French scientists was made. In this case a more physiological approach was adopted and the French authors used earlier chemical analyses in their work. These were not always reliable and the existence of hydrochloric acid in the gastric juice was denied on the authority of Berzelius who ascribed the acidity of this fluid to the presence of lactic acid. On this point a controversy developed; the French authors challenged the accuracy of

Prout's analytical procedure and the latter replied by repeating in detail his description of the method, claiming that it was 'quite perfect'. He did not deny that other acids besides the hydrochloric might be present in gastric fluid, but insisted that hydrochloric acid was the main cause of acidity during digestion. (Organic acids might be present in the stomach as a result of the oxidation of food matter.)

By this time Prout was also well known for his views on the ultimate constitution of matter and his attempts to apply the atomic theory to the results of organic analyses. He was fascinated by the apparently simple numerical relationships which obtained between the 'molecular weights' of so many of the chemical substances found in the living body and guided by these ideas he went on to work out a system of chemical changes thought to occur during digestion. These were then included in his Bridgewater Treatise on natural theology entitled, *Chemistry, Meteorology and the Function of Digestion*, published in 1834. In this book Prout spoke of *weak* compounds, having a complex structure leading to unstable molecules which were readily broken down into simpler, more stable compounds, which were regarded by contrast as *strong*. This conversion was usually found to be brought about by the addition of water to the weak compounds during the early stages of digestion and was termed *reduction*. Later, during respiration, some of the excess water was thought to be removed and the alimentary substances were prepared for their introduction into the blood stream by a further process called *completion*. This whole series of changes constituted the process of animalisation and was thought to take place under the influence of 'organic agents', whose precise nature was unknown.

Prout suggested that the digestive system as a whole was arranged upon galvanic principles, with the stomach as the positive pole and the liver as the negative. Common salt in the body fluids was thought to be electrolysed so as to release hydrochloric acid in the stomach and soda in the bile—a neat way of accounting for the observed composition of these fluids and at the same time explaining the body's need for a continuous supply of salt. Prout was also important for his suggested classification of foods into three main groups, viz., saccharinous (carbohydrates), oleaginous (fats), and albuminous (proteins), which, together with water, were regarded as the four alimentary principles.

Amongst all the work on digestion in the nineteenth century the most dramatic was that of the American physician William Beaumont, who, by his fortuitous involvement with the victim of a serious gunshot wound, was enabled to study the behaviour of the living stomach and the nature of its secretions. Alexis St Martin, a French Canadian, was wounded in an accident when a shotgun exploded behind him at point-blank range. The charge passed right through his body and it was only due to his own fitness and the expert attention which he received from Beaumont that he survived the accident. However a gastric fistula which refused to heal remained and through this the physician could obtain access to the patient's stomach. Between May 1825 and November 1833 Beaumont supported St Martin in his own home in exchange for the privilege of performing upon him a long series of carefully planned experiments in which he observed the sources of secretion of the gastric juices and extracted samples from the stomach for chemical examination. Beaumont stated his observations objectively, leaving the reader to draw his own conclusions. He gave an accurate and complete account of the nature and properties of gastric fluid, confirming Prout's observation that the important acid present is hydrochloric acid. He noted the great influence of exercise, the weather and the emotions on the secretion of gastric fluid; he studied the motions of the stomach and examined the digestibility of many items of the normal diet. It is interesting to note Beaumont's conclusion that gastric juice 'contains free muriatic (i.e. hydrochloric) acid and *some other active chemical principles*'. This was an early indication of the presence of an enzyme in the gastric fluid, later to be confirmed as 'pepsin' by Theodor Schwann in 1836.

In this chapter we have considered briefly some of the chemical work done by early nineteenth century physicians who were interested in animal chemistry as it applied to medicine. The problems of biochemistry are complex and these men were not equipped to solve them; nevertheless, they worked in the belief that such problems were *capable* of solution by chemical and physical methods. They tackled those parts of the subject which were susceptible to investigation by the available techniques and they made some important, if limited, discoveries. Most animal chemists recognised the mystery surrounding the chemistry of life and did not rule out the possibility of the presence of vital forces or agents. Yet, in so far as chemistry was found to be applicable to the vital functions, the animal chemists were confident that it was the same chemistry as obtained amongst mineral substances. On the other hand, many physiologists shunned animal chemistry in the belief that it was incapable, in the last analysis, of accounting for the mechanisms of the vital functions. In the next chapter we shall attempt to find out more about the vitalist's outlook.

Chapter 5

Debates about vitalism

SINCE ANIMAL AND PLANT organisms were so complex and their functions proved so difficult to study it was widely accepted that they must involve the presence of some unknown principle or force other than the simple physico-chemical ones. The functions of life were peculiar, it was argued, in that they needed some form of vitality. This concept, which can be traced back to ancient times, had appeared in the work of Paracelsus and van Helmont as the 'archeus'—an indwelling spirit governing the vital functions of digestion, respiration, nutrition and so on. By the eighteenth century this notion had become modified, but the concepts of vital forces, principles or laws were still very popular as a means of 'explaining' otherwise inexplicable life-functions.

The first systematic treatment of the concept of vital force appeared in the opening paper of the first issue of *Archiv für Physiologie*, published by J. C. Reil in 1795 at the University of Hallé. Reil based his discussion firmly on experimental evidence and avoided all flights of fantasy, so common among vitalists. He emphasised that the phenomena of animal life must be deducible from material considerations and thought that life arose from the particular 'mixture' of elements and compounds in organised matter (i.e. their chemical complexity). Chemical analysis, which could be expected to elucidate the form of organised matter, would thus become ever more important to theoretical and practical medicine.

Reil felt that the term 'force' was too vague and preferred to speak of property of matter instead. Since he believed that vital force arose out of the special organisation of living matter he thought that with increasing chemical knowledge we might hope to arrive at a greater degree of precision with respect to the concept. He was convinced that the various manifestations of life would in the end be explained in chemical terms and thus it would be possible to dispense with the hazy idea of vital force. This outlook was ultimately adopted by many animal chemists, but Reil's rational approach did not appeal to most vitalists and the majority of animal chemists and physiologists continued to speak of vital force throughout the nineteenth century.

Chemistry and Vitalism

Berzelius undoubtedly had an important influence in maintaining the vitalistic concept in the thought of the animal chemists. He considered that vital force was something which did not belong to the ordinary sphere of inorganic elements and compounds. It was basically different from properties such as gravity, impermeability, electrical polarity and so on. What it was and how it came into being or disappeared was simply not understood. The chemical elements in organised matter acted under its influence. Berzelius did not substantially change his views throughout his long career and other animal chemists, following his lead, often found it necessary to adopt the vitalistic concept. It was generally understood however, that advancing chemical knowledge would consistently reduce the areas covered by the concept and animal chemists therefore worked in an air of expectancy, looking forward confidently to the time when organised substances could not only be analysed but also synthesised in the laboratory. The most difficult problem was to understand how the complex chemical changes known to occur in the living body could go on at such moderate temperatures. It was in an effort to account for this that the vital force was often invoked.

William Prout argued that since the same chemical elements were present in organised matter as in mineral substances, chemistry had an important part to play in the natural functions. He suggested that there were peculiar substances which existed in all organised matter and governed its behaviour. These he called 'organic agents' or 'intelligent agents' and he regarded them as essential for the synthesis of life products. Thus, a compound such as sugar could not be synthesised in the laboratory because the chemist did not understand the action of the necessary organic agents. Prout thought that the most striking feature of organised matter was its great diversity which was produced by the vast number of organic agents causing the same chemical elements to combine in an almost infinite number of ways. It was only as enzymes such as ptyalin, pepsin and diastase came to light that satisfactory chemical explanations began to emerge. Prout's organic agents, responsible for vitalising inanimate matter were after all chemical compounds; perhaps the vital force or vital principles would turn out to be recognisable in the same way.

In 1800 chemistry and physics could explain very little in the field of physiology. Certainly the work of Lavoisier had shown that respiration was a chemical process connected with digestion and the evolution of animal heat, but physiologists could not be expected to accept this as proof that physical science could provide viable accounts of all other life-functions. To assert this at the beginning of the nineteenth century required an act of faith which most physiologists were unwilling or unable to make. As it became clear that the four main elements in organised matter were the same as those present in the atmosphere there was seen to be a balance between living organisms and the air. Nevertheless, most chemists and physiologists accepted the idea that the common elements came under the organising influence of vital force when they were incorporated into living matter.

Fourcroy tacitly assumed the presence of vital forces in the body from the difficulty of synthesising organic compounds. He pointed out that it was only possible to prepare the simplest animal substances by means of the ordinary techniques of chemistry in the laboratory and implied the existence of some vitalising influence in living organisms. Nevertheless, Fourcroy was not truly a vitalist since he thought that it would ultimately be possible to provide a physico-chemical explanation of all life-functions. He suggested that chemists might begin by comparing the analyses of vegetable and animal matter in an attempt to determine what happened during 'animalisation', the fundamental vital change which was thought to occur during digestion. It seemed that this change was brought about by the addition of hydrogen and azote (nitrogen)—a simple chemical process—yet to bring about such a change in the laboratory proved very difficult and even impossible. In nature chemical forces appeared to work in direct opposition to the vital ones and when life was extinguished the chemical forces of dissolution began to take over, decomposing by oxidation those compounds which had been synthesised under the influence of the vital force.

The treatment of Vitalism amongst physiologists

In the first half of the nineteenth century one of the most popular textbooks of physiology was that of J. F. Blumenbach, the German physiologist best remembered for his studies in the comparative anatomy of man. This book, originally written in Latin under the title *Institutiones Physiologicae*, was translated into German, Dutch, English, French, Spanish and Russian. Blumenbach taught that the body was composed of solids endowed with vitality and thus capable of receiving and reacting to stimuli. Vitality was thought to be easy to recognise by the property of irritability, although it was difficult to define. Yet it was clear that the living animal was endowed with peculiar properties which governed its behaviour and which persisted for some time even after death. Such properties did not seem to be accountable in physico-chemical terms, although they were often associated with physical and chemical processes. Blumenbach classified the vital powers into three groups under the headings (i) organic formation and increase, (ii) motion in the parts when formed, (iii) sensation. This, he pointed out, was the order in which the vital powers appeared and developed during gestation and in the period immediately following birth. The first of the three vital powers was thought by Blumenbach to be the most important and was called the 'nisus formativus'. It was the force which generated the organs of the body, preserved and nourished it during life and reproduced as far as possible, any part which was mutilated. This formative power could be recognised in the solid parts of all living things whether plant or animal, but its existence in the fluid parts was less certain. The only animal fluid thought to be endowed with vital force was the blood and even in this case, Blumenbach suggested that the apparent incorruptibility of blood was due more to the perpetual chemical changes which it underwent during respiration than to any vital power. Thus, whenever a chemical explanation seemed possible, even the physiologist tended to adopt it in preference to the loosely defined vitalistic concept.

Blumenbach also recognised that the vital functions were closely associated with mental activity, although the former could continue without intermission whilst the latter needed periods of rest. Amongst the vital functions he included digestion, circulation, respiration, exhalation, absorption, secretion, nutrition and calorification, all centred upon the heart. Mental functions, which were thought to include sensation, voluntary muscular action and the nervous functions in general, were centred in the brain.

In 1795 an American edition of Blumenbach's work appeared, prepared by Charles Caldwell, who believed quite firmly that the functions of life could not be explained in chemical terms. It was not possible to produce fat, skin, muscle, hair, feathers, bone or any other parts of the living body in the

laboratory, nor to bring about artificially even the most fundamental effects of the vital functions. Such products and reactions could only be brought about under the influence of the
vital powers and in Caldwell's extreme view, chemistry *only*
invaded the body when life had left it. Chemical forces operated
in opposition to vital ones, destroying by oxidation, fermentation and putrefaction, the substances which had been built up
by the vital powers. It often seemed that the most important
function of the vital powers in the living body was to resist
the physical and chemical forces of oxidation and dissolution.
Indeed, many physiologists regarded this as the most characteristic property of the vital powers. It was known that a living
muscle for example, would sustain a weight which was sufficient
to break it when removed from the body and this seemed to
show that life gave the muscle much greater strength. Examples
such as this lent a validity to the concept of vital force which
it was very difficult for the chemist, with his imperfect knowledge, to counter. Caldwell therefore chose to explain life by
means of a material vital principle, rejecting all the chemical
and chemico-physiological accounts.

Few were prepared to adopt such extreme views and in fact
not all physiologists were vitalists. For example, John Bostock
the Liverpool doctor, rejected the concept of vital force both
as a general description of life functions and as a specific
explanation of their modes of action. He held that it was
unreasonable to use the notion of a single principle to cover
all the diverse processes of life, since this would be to generalise
specific differences and reduce them all to one class of phenomena simply because they could not be fitted into any other
class. So Bostock rejected vital force as a cover for ignorance
and was prepared to attempt to explain life in terms of the
physico-chemical mechanisms of functions such as digestion,
respiration, excretion and so on. He criticised those who
offered explanations which went beyond the limits of experimental observation. Speaking of digestion, Bostock wrote in

1826, '... we have only an imperfect acquaintance with the successive steps in the operation, and we are, in a great measure, ignorant of the nature of the agents by which it is effected.' Bostock was convinced nevertheless, that the full explanation of digestion would ultimately be given in chemical terms. There was a great deal of interest in the chemistry and physiology of digestion at this time. William Prout had announced the discovery of hydrochloric acid as an essential constituent of the gastric juice in 1824 and in the following year the Academie of Sciences in Paris had offered a prize for the best essay to be submitted on the function of digestion. As we have already seen, two attempts to secure this prize were made, one by the French workers Leuret and Lassaigne and the other by the German scientists, Tiedmann and Gmelin. Both essays were published in 1825 and the German workers confirmed Prout's discovery. These studies of the digestive process, admirable though they were, still left much detail unexplained; it was the very complexity of the process, as with other life-functions, which left the nineteenth century chemist powerless to provide really full explanations. Bostock supported the rational approach to all such problems, but he knew that some animal functions, such as respiration and the action of the heart could be maintained mechanically without producing life—clearly, it seemed, something else was also present in life processes.

It has frequently been assumed that the concept of vital force was discredited amongst chemists by Wöhler's preparation in 1828 of urea $(NH_2.CO.NH_2)$ from ammonium cyanate (NH_4CNO). In recent years however, this view has been seriously challenged. Wöhler himself was not especially anxious to claim this reaction as an example of the laboratory synthesis of a product of vital activity. It appears that he was far more interested in the idea that urea and ammonium cynate were isomeric and that he had added a new example to the cases of isomerism already discovered by Berzelius. The reaction had no impact at all on Berzelius' vitalistic views, although he too accepted it as another example of isomerism. Of course, there is a sense in which the formation of urea from ammonium cyanate might be regarded as a re-arrangement of the elements in this compound, rather than a total synthesis. The ammonium cyanate itself could be obtained from animal sources and so it could be argued that the influence of vital force was already active in this compound, before its transformation into urea. On the other hand there were those who did not regard excretory products like urea as part of 'vital chemistry', even though they had been formed in the body.

Thus, Xavier Bichat, the French physiologist, thought that life functions were so complex that they could only be understood in terms of a 'totality of vital properties', but these were operative only within the living body. The chemistry of body fluids such as urine, saliva and bile, when removed from the body, was merely the 'dead anatomy' of these fluids which no longer had any claim to vitality. Müller, the leading German physiologist, also doubted whether urea should be considered as 'organised' since it was an excretory product and not a component of the *living* body. The difficulty was to decide where to draw the line distinguishing 'organised matter' imbued with vital force from inanimate matters which also formed components of the living body.

Vital force and the nervous system

Berzelius was in no doubt that all the components of the body should be examined *chemically* and that chemistry and physics should be introduced as explanatory devices when dealing with all animal functions. To reject these in favour of vitalism would be to describe the animal functions without explaining them. But, though life processes were undoubtedly chemical in nature, they needed a special driving force which was to be found in the nervous system. In his *View of the*

Progress and Present State of Animal Chemistry, Berzelius wrote, 'This unknown cause of the phenomena of life is principally lodged in ... the nervous system, the very operation of which it constitutes.' Unfortunately chemistry had nothing to say about the way in which the nervous system functioned and since he thought it was here that vital force had its origins, Berzelius believed that animal chemistry would always be at a disadvantage. This unknown force entered into every vital process and it seemed that the highest knowledge the chemist might ever attain would be that of the chemical constitution of the animal products themselves. The mechanisms of the functions by which they were produced might always elude him. This approach to animal chemistry always coloured Berzelius' outlook and it accounts for the fact that he was more concerned with the chemical analysis of animal substances than with the study of animal functions. Other animal chemists, notably Prout and Liebig, viewed the future possibilities of the subject more optimistically.

Nevertheless, the suggestion that the vital force might be found in the nervous system was an interesting one, for it was clear that the possession of such a system was a most important distinguishing feature of all animals and the activities of the nervous system undoubtedly governed the vital functions. Severe damage to the nervous system could be fatal and it seemed fair to assume that vitality resided in the nerves. Galvani's discovery of electric currents in the frog in 1791, led to the idea that it was this mysterious electric fluid which, flowing in the nerves, gave rise to vital force. Galvani thought that he had at last discovered the secret of 'animal spirits' in the electric discharges between the nerves and the muscles. Could this in fact be the solution to the problem of vital force? The theory of galvanism, or animal electricity was summarised in the mid-nineteenth century by Emil du Bois Reymond, a German physiologist despite his French name. Reymond held that there was a peculiar form of electricity present in the animal body, secreted by the brain and distributed throughout the nervous system. The muscles, acting as receivers of this animal electricity, contracted when it was discharged through them from the interior to the surface via the nerves. All muscular motion was stimulated by electric impulses in this way. There was a general interest in the nervous system by anatomists and physiologists in the nineteenth century. Many experiments were performed to determine the action of the central nervous system in controlling the involuntary functions such as heart-rate and vital functions like digestion and respiration.

Amongst vitalists who worked in this field in the early years of the century must be mentioned A. P. Wilson Philip, physician at Worcester. From his experiments with rabbits and frogs Philip concluded that there were three levels of vital powers, (i) muscular, (ii) nervous and (iii) sensory. He found that both muscular and nervous activity was destroyed when the sensory system was destroyed, whilst chemical or electrical stimuli applied to the nerves exerted a greater effect on involuntary muscular movement than did mechanical stimuli. Philip used his results to devise galvanic treatments consisting of electric shocks and weak electric currents which he applied successfully to patients suffering from asthma and various forms of paralysis.

Philip also demonstrated that the muscles of voluntary motion were stimulated only by the specific parts of the brain and spinal marrow from which nerves originate, whereas the vital organs such as the heart, stomach and lungs were stimulated by all areas of the brain. This was found to be due to the fact that the vital organs are supplied by ganglionic nerves which derive their origins from many sources in the brain and spinal marrow. The control of the vital organs is, as a result, much more complex and they are affected by widely different mental influences. The secretions were also found to be derived from the blood under the influence of ganglionic nerves and analogies which Everard Home described between the structures of the ganglia and the electric organs of fishes such as the torpedo, supported the view that nervous impulses were electrical.

Home also suggested that the production of animal heat, another important vital function, might be controlled by electric currents in the ganglia and this idea was investigated by a fellow surgeon and colleague in the Animal Chemistry Club, B. C. Brodie, in a series of experiments on rabbits and dogs. In some of these experiments Brodie removed the head from the animal, taking care to prevent undue loss of blood. He found that the heart continued to beat for some time and he ventilated the lungs mechanically just so long as the heart-beat persisted—as long as two hours in one case. Brodie kept a constant check on the temperature of the body and on the rates of respiration and heart beat. In all cases however, he found that though the rate of artificial respiration was maintained equal to that of normal breathing and the blood continued to circulate whilst undergoing all the changes observed in life, the temperature of the dead animal always fell. In fact the rate of fall was found to be more rapid than that of a similar dead animal whose body was allowed to cool in the natural way. It followed that the air which had passed through the lungs had served to cool the body and no heat had been generated by the artificial respiration.

In a second series of experiments Brodie killed the animal by means of poisons which had a less drastic disturbing effect on the central nervous system than the removal of the head. Artificial respiration cooled the body as before, but in an experiment on an ass, Brodie found that if the artificial respiration were continued until the effects of the poison had worn off, the animal recovered sensibility and its capacity to generate heat returned in proportion to the degree of restoration of its nervous energies. From this it seemed clear that the generation of animal heat was also governed by the nervous system.

A different method of demonstrating the same point was chosen by Everard Home who divided the nerves to one of the two growing antlers of a stag. Measuring the temperature of the two antlers Home found that the un-nerved one was initially at a lower temperature, but that this rose daily until after about a week it had regained the same temperature as the other. The stag was then killed and it was found when the antlers were examined that although the old nerve endings had not rejoined, new connections had formed. Home concluded that the temperature of parts of the body was directly dependent upon the nerves and when their influence was removed the temperature fell.

Relationships between heat and life were also investigated by Charles Chossat, a French physiologist, who showed that birds could be made to cease all their natural functions if their bodies were cooled sufficiently. It seemed that heat as well as electricity was a stimulus capable of arousing the vital force; clearly there was a close relationship between vital and physical phenomena. Several nineteenth century physiologists and physicists considered this relationship, drawing analogies between vital force and gravity, electricity, magnetism, light, heat, chemical changes and so on. It was suggested that vital force should be regarded as the basis of physiology, just as affinity was the basis of chemistry and mutual attraction that of physics. Physiologists might then set out to determine the laws of vitality in the same way that chemists and physicists had established natural laws in their own branches of study.

Fermentation: Chemical or Vital?

Turning to the study of fermentation and putrefaction we find that the chemical and vitalistic arguments are in direct conflict. Lavoisier had regarded these two vital processes as chemical reactions depending on the decomposition of water.

The oxygen of the water was said to combine with part of the carbon of sugar to form carbon dioxide, whilst the hydrogen combined with another part of the carbon to yield alcohol. Sugar was considered to be a loose compound readily decomposed by the weak forces exerted when a small quantity of yeast paste was mixed with the aqueous solution. Putrefaction was considered to be a form of fermentation in which hydrogen was evolved and stable compounds such as ammonia, phosphine and hydrogen sulphide were formed.

Now, although these were undoubtedly chemical reactions, it was realised by some chemists that yeast was a living substance and that fermentation was brought about by the action of micro-organisms. Thus fermentation was the result of a vital process. In 1836 Cagniard de la Tour, French engineer and physicist, observed the reproduction of yeast which he saw to be a mass of minute organisms multiplying by budding. Fermentation occurred at the same time as a consequence of this life-process—the products were chemical but the process leading to their formation was 'vital'. Theodor Schwann showed that air heated to 360°C. became incapable of initiating fermentation or putrefaction; it had become sterile and the micro-organisms which it would normally contain had been destroyed.

Schwann had been working on alcoholic fermentation and had shown that yeast was growing during the process. He pointed out that the changes occurring in the yeast cell were some of the best known and simplest examples of the processes which go on in every living cell. It was in 1839 that Schwann finally published his views on the cellular nature of all living matter and characteristically his book began with a discussion of the controversy between the mechanistic and vitalistic views of life processes. Schwann decided in favour of the mechanists but concluded that the cause of nutrition and growth in an organism was to be found in its individual particles—its cells. The force displayed by the living cells came into operation as a result of their interactions with their environment within the living organism as a whole. Cells could draw materials from their environment and change them chemically, producing what Schwann called a 'metabolic force'. In this way the concept of vital force was placed firmly within the living cell and shown as a consequence of the chemical changes taking place there.

E. Mitscherlich, professor of chemistry at Berlin, was one of the first chemists to accept that yeast is a living organism consisting of cells. In an experiment in which a suspension of yeast cells was separated from a solution of sugar by a layer of filter paper, Mitscherlich was able to show that *contact* with yeast cells was necessary for fermentation to occur. The sugar which passed through the filter paper was fermented, but no fermentation went on in the main part of the sugar solution which was isolated from the yeast cells. This need for contact seemed to emphasise the similarity between the process of fermentation and catalysis. Berzelius, who thought that catalysis itself depended upon contact between the catalyst and the reactants accepted this similarity and insisted that yeast simply acted like an amorphous inorganic catalyst.

Liebig, whose authority as an animal chemist rivalled that of Berzelius, agreed with Lavoisier in thinking that fermentation and putrefaction were purely chemical processes, although he suggested that they were initiated by the oxygen of the air and that once begun they would continue in the absence of air. They were maintained by the disruptive molecular motions of ferments which broke up the molecules of sugar by mechanical means. There appeared to be a direct analogy between this molecular motion and that which could be produced by physical forces such as chemical affinity, electric and magnetic forces, cohesion and so on. Whatever the nature of the force acting, the final result was the same whenever the resistance to the disruptive force was insufficient—the larger molecules were broken down into smaller simpler ones. Liebig compared the action of the ferments to the mechanical disturbance caused when delicate compounds such as chlorine dioxide or silver fulminate were struck with a hammer. The blows would be enough to cause the explosive decomposition of such compounds and ferments in the same way caused the decomposition of sugar. Speaking of the analogy between the action of the ferments and that of a catalyst, Liebig suggested that the latter overcame 'inertia' between the reactants and stimulated an otherwise stable chemical system into action. The catalyst merely accelerated a process which normally occurred very slowly and led to the formation of more stable products. In the same way a ferment accelerated the decomposition of organised matter. Carbon dioxide, the most stable compound of carbon, was always found as an end product of fermentation, whilst in putrefaction the products always included also ammonia, phosphine, hydrogen sulphide etc., the most stable compounds of these elements. The process, for Liebig, was entirely chemical and he thought that insoluble yeast found suspended in the fermented liquor was formed by oxidation of a soluble nitrogenous substance, gluten, present in the liquid undergoing fermentation. This process was therefore fundamentally an intramolecular oxidation-reduction reaction. Thus,

sugar → carbon dioxide   + alcohol
         (oxidised product)   (reduced product)

The weakest point in Liebig's theory was that it did not account for the *specific* nature of the ferments. Yeast was known to be capable of causing fermentative and putrefactive changes in other substances, but it was not the only material with this capacity, for it was found that in certain cases muscle tissue, urine, egg-white, fish-glue, cheese, blood and other organised substances could all, when in a state of putrefaction, initiate similar fermentative and putrefactive changes. Each of these substances could bring about only certain changes and in each case the products of the reaction were different. Liebig recognised this but made no attempt to explain why one form of ferment worked in some cases whilst a different one was required in others and each ferment gave rise to specific products.

In contrast to the purely chemical view of fermentation, Louis Pasteur supported the vitalist explanation. In 1860 he asserted that the process was inseparable from vital activity and declared that there could be no fermentation of any kind without the presence of living cells. About the same time, van den Broek professor of chemistry at the Military School in Utrecht, showed that grape-juice squeezed out under mercury without coming into contact with the air, did not change for months. He concluded that fermentation depended upon the vegetative growth of yeast cells, stimulated by some agents in the air which could be destroyed or removed by heat or filtration through cotton-wool. He also found that if animal fluids such as blood, urine or bile were collected in sterile vessels over mercury or even in contact with air provided that it had been heated or filtered, they did not putrefy even if kept for long periods.

In the following year Pasteur showed that yeast and other micro-organisms which cause fermentation could grow in the complete absence of air, provided that they were in contact with sugar or acids or other matter from which they could extract oxygen. He proposed the names *aerobic* and *anaerobic* for organisms capable of living with and without atmospheric oxygen respectively. In his *Études sur la Bière*, published in 1876, he stated his famous conclusion, '*La fermentation est la consequénce de la vie sans air.*' Pasteur's work was in direct opposition to the purely chemical accounts of fermentation favoured by Berzelius and Liebig. It elicited a long critical Memoir from Liebig in which he pointed out that the assumption that fermentation is always the result of vital activity in living cells could not explain the action of diastase, pepsin and emulsin. In fact he claimed that the vital activity resulted only in the production of such unorganised ferments in the living cells. Experiments had already been made which indicated this and the theory that fermentation is brought about by unorganised ferments (i.e. enzymes) fabricated in the living cells was shortly to be confirmed by Moritz Traube, a pupil of Liebig. Traube was able to confirm Pasteur's observation that yeast would grow in sugar solution in the absence of oxygen, but he found that the growth was much more rapid when atmospheric oxygen was present. He concluded that the growth was independent of the fermentation which was caused by a chemical ferment in the yeast. In 1858 Traube called for research into the isolation of the chemical ferments produced by different infusoria or moulds involved in fermentation or putrefaction.

Many experiments were done in attempts to demonstrate that living cells were necessary for fermentation to occur, but the results tended to be confusing because in general too little care was taken to ensure that the air was excluded from the apparatus. Helmholtz, for example, boiled infusions of meat and then separated the liquor by a bladder from another putrefying infusion. He found however that the boiled preparation also began to putrefy and, thinking that the putrefaction had spread through the bladder, he concluded that the process was not dependent upon germ-life. On the other hand, early experiments had indicated that if yeast were triturated until all the cell structure had been destroyed, the product would not ferment glucose even after two days.

It was only in 1897 that Buchner was able to show that alcoholic fermentation is caused by a non-organised ferment. Buchner crushed yeast cells by triturating them with sand and kieselghur. The paste so formed was then pressed through a cloth bag to obtain a liquid free from yeast cells which was still capable of causing alcoholic fermentation. The activity of this solution was found to persist even after it had been evaporated to dryness at 30-35°C. and then reconstituted by adding distilled water. Such unorganised ferments, or enzymes had already been recognised in various stages of the digestive process by Berzelius and others although the name had only been introduced in 1878 by Wilhelm Kühne, professor of physiology at Heidelberg. The enzyme in Buchner's extract of yeast was recognised as 'zymase' by Béchamp. Liebig ultimately became convinced of the organised nature of yeast itself, but he stuck to his theory of molecular agitation communicated to the sugar by the ferment. His views on this, as on some other aspects of vital chemistry, were not confirmed by later work.

Summary

As animal chemists and physiologists strove to elucidate the structures and functions of living things the debate about vitalism continued to play an important part in determining the direction in which conceptions about the nature of life were to develop. At first there had been the concept of a vital principle, occult, non-material, a purposive entity governing the vital functions. Then the very complexity and variety of life began to impress physiologists such as Bichat for example, who insisted on the importance of variability in the laws of vitality. Vital phenomena are marked by an irregularity which distinguishes them from the regular physico-chemical pheno-mena of nature, he said. In fact, Bichat chose to interpret this variability and irregularity of behaviour as a capacity in living things to *evade* the accepted laws of nature. Animals for example, seem to disobey or disregard Newton's law of cool-ing by the production of just enough heat to maintain a reason-ably steady body temperature. In the early years of the nineteenth century it had been incumbent upon the animal chemists to show that the laws of physics and chemistry applied to living things, but by the end of the century the emphasis had so changed that it had become necessary for the physiologists to show that they did not. In this respect at least animal chemistry had proved itself successful.

In another approach to the problem of vitality it was suggested that organised matter in some way contained the vital force within its material constituents. The components of such matter were found to be so complex that it seemed possible that their very complexity might give rise to properties different in kind from those of simpler inorganic or mineral substances. The discovery of enzymes, vitamins and hormones might have lent support to this view, but these did not come before animal analysis in the hands of Berzelius and Liebig had shown that the same laws of chemical composition applied to the organic as to the inorganic compounds.

Liebig had tried to view the vital force on a level which was strictly comparable with the forces of gravity, electricity, magnetism, chemical affinity, etc., but whilst the vital force was thought to obey similar laws to the physico-chemical forces it was at all times capable of over-riding them. Vital force was capable of causing and sustaining growth; it also entered into the balance sheet of the conservation of energy along with the other forces. Unfortunately Liebig's ideas on this topic aroused little response because animal chemists were intent on showing the identity between organic reactions in the body and in the laboratory. Most were therefore unwilling to turn back and try to trace distinctions between normal chemical bonds and 'physiological bonds' involving a vital force component, unless this could be shown to be necessary.

Fermentation studies seemed to indicate that there were organic chemical reactions which could only occur in the presence of living organisms. When it was found that the products of such reactions were often optically active, this seemed to be evidence that there was something very special about vital activity. Experiments showed that alcoholic fermentation by yeast was accompanied by the growth and multiplication of the living yeast cells and this led to the further requirement that to elucidate vital chemistry it would be necessary to study the reactions which occur within the living cell.

Claude Bernard realised the importance of the cell and especially of the interchange which occurs all the time between the contents of the cell and the fluids, derived from the blood, which surround it. Vital chemistry, he suggested, depended upon the constancy of this 'internal environment'. Bernard considered that the animal body functioned according to unchanging physico-chemical laws, but the *processes* which were involved were so complex that they should be regarded as specialised physiological ones. In this way life-functions could be given their special character without departing from the principles of physics and chemistry. The specialism arose, not from special laws, substances or forces, but merely from complexity. Only when he came to discuss embryology did Bernard's faith in the purely physico-chemical approach fail him. Here he assumed a 'creative idea' which was present in every living germ and was able to unfold and show itself through organisation. Bernard's 'ideas', suggested long before the sciences of genetics and sub-cellular chemistry had been developed, might be considered as precursors of our chromosomes and genes.

The concept of vitalism is an attractive means of covering ignorance in the absence of detailed knowledge and experimental fact. It is a 'blanket term' by which much can be apparently explained and its more fervent adherents have always linked it with theological ideas, seeing through its use the expression of God in nature. This became very strong amongst exponents of Naturphilosophie in Germany and in England the study of Natural Theology is exemplified by William Paley's *'Natural Theology; or Evidence of the Existence and Attributes of the Deity collected from the Appearances of Nature'*, published in 1802, and by the famous

Bridgewater Treatises of the early 1830's. These works set out to demonstrate the presence of God in the Natural World through a study of His works. As a result of such connections vitalism came to have an emotive connotation which was sometimes expressed in vituperative, emotional terms and led to many unnecessary, verbose and fruitless controversies.

Chapter 6
Animal Chemistry in France

IN FRANCE after the Revolution there was a determined attempt to reshape every facet of life on scientific lines. A new calendar was introduced and new standards of measurement, based upon experimental observations were defined. Amongst a whole array of new organisations, civil, political and military, those established for the purpose of fostering scientific research took pride of place. Napoleon himself displayed a keen and active interest in scientific matters and was very proud of his membership of the scientific class of the *Institut de France* (renamed *Académie des Sciences* in 1816). He had gained this distinction on his own merits before he rose to power and he always gave special consideration to all requests for support for genuine scientific research. Amongst famous French scientists and mathematicians, Napoleon befriended Laplace, Lagrange and Berthollet, together with many others. At that time chemistry was the science which seemed to offer the highest rewards from its applications to arts and manufactures. Consequently chemical research became important in early nineteenth century France. French chemists, mainly in Paris, led the world both in numbers and in the quality of their work.

The natural sciences, including the chemistry of organised bodies, were studied at the *Muséum d'Histoire Naturelle* which had grown out of the *Jardin du Roi* of pre-Revolutionary days. In 1784, Fourcroy was appointed professor of chemistry at the *Jardin du Roi*, on the death of P. J. Macquer who had held this post since 1771. Fourcroy was chosen largely as a result of the high reputation which he had established as a lecturer. He remained in this post until his death in 1809, his reputation as a chemist and as a teacher of chemistry growing all the time. At the *Muséum* pharmaceutical chemistry was important and the chemists also taught those branches of their subject which were ancillary to geology and to animal and vegetable analysis. The chair of general chemistry held by Fourcroy, and later by Gay Lussac, was also complemented by another chair of applied chemistry which was filled by Vauquelin and later by Chevreul.

In December 1794 a new teaching establishment was set up in Paris under the title of *École Polytechnique*. This was founded for the purpose of providing a common training for civil and military engineers as well as for geographers, naval architects and those who would go on to teach mathematics and the physical sciences. The founders of the *École* included a number of influential chemists amongst whom Fourcroy was prominent. Both he and Berthollet lectured there, as did Vauquelin, Chaptal and Guyton de Morveau. This led to an undue emphasis on chemistry which consequently became one of the principal subjects in the curriculum of the *École*. The study of this subject also included a high proportion of student practical work—unusual in those days owing to the expense of apparatus and chemicals. The laboratories of the *École* were equipped from many sources, including gifts from Napoleon himself who was especially generous in the donation of large quantities of expensive mercury.

Amongst the influential French chemists of this time, C. L. Berthollet was the first to declare himself in favour of Lavoisier's anti-phlogistic theory, although he had earlier defended the phlogiston theory. Berthollet announced his adoption of Lavoisier's new theory in April 1785 and at the same time he declared that phlogiston was a useless hypothesis. Befriended by Napoleon, Berthollet took an active interest in the organisation of French industry after the Revolution. He introduced the method of bleaching cloth with chlorine and his activities in connection with improvements in the manufacture of saltpetre, steel etc., made France largely independent of foreign imports at that time.

Berthollet's main interests were in research; he was a careful, conscientious practical worker, but his lectures often proved too difficult for his students. He included in them some discussion of vegetable and animal matters as compounds of carbon and hydrogen and from the observation that oxalic acid could be obtained from both vegetable and animal substances when treated with nitric acid, Berthollet recognised a common basis for all organised matter. He also suggested that in the case of vegetable substances there was a correlation between the quantity of oxalic acid formed and the nutritional value of the plant.

Berthollet thought that the chief chemical difference between vegetable and animal matters was connected with the proportions of azote which were found to be very much greater in animal substances. The latter also yielded phosphoric acid and an excess of carbon which remained as charcoal after destructive distillation. In this way animal matters were seen to be more complex than vegetable substances and Berthollet concluded that in general they probably contained a substance similar to sugar, a kind of oil, phosphoric acid combined with lime, phlogisticated air or 'mofette', and some fixed air or carbon dioxide. He found that when silk, an important product of French industry, was heated with nitric acid, oxalic acid was formed and nitrogen released. During dry distillation and putrefaction of animal substances, on the other hand, much of the nitrogen was found to combine with inflammable gas (hydrogen), to form ammonia. When animal muscle was subjected to dry distillation a peculiar acid substance was evolved which Berthollet named 'zoonic acid' in 1797. This was shown to be merely an impure form of acetic acid by Thenard in 1801.

Berthollet was also important for his part in sponsoring chemical and physical research in his private laboratory at Arcueil, near Paris. Laplace also lived at Arcueil and between them these two fostered such an interest amongst younger

French scientists that by 1807 there had grown up a group strong enough to form a Society with its own Journal. Most of the members of this Society were connected through their work in one or other of the scientific institutions in Paris, but they met at Arcueil to pursue their experimental researches free from public engagements or teaching commitments. They worked under the general direction of Berthollet. The group included Gay Lussac, Thenard, Biot, Chaptal and others, whilst foreign visitors, like Humboldt, frequently came to Arcueil at Berthollet's invitation. Volta, Blagden, William Herschel and Oersted were early visitors there; Davy and Berzelius also paid visits to Berthollet's laboratory. Whilst chemistry was studied with the natural sciences at the *Muséum* in Paris, at Arcueil it was considered as a branch of the physical sciences.

Gay Lussac had come to Paris in November 1794 and had begun to study for entrance to the *École Polytechnique* which he gained three years later, at the age of nineteen, having passed the entrance examinations with distinction. After graduating in 1801 he went to work in Berthollet's laboratory at the *École*. Then in 1804, Thenard, who had been Fourcroy's demonstrator at the *École* since 1798, was appointed professor of chemistry at the *Collège de France* and Gay Lussac became demonstrator to Fourcroy in his place. Thenard's work was closely allied to that of Fourcroy and Vauquelin; he concerned himself largely with the chemistry of natural substances. Gay Lussac, on the other hand had received the rigorous mathematical training of the *École* and therefore turned his attention towards physics and physical chemistry. The influence of Berthollet upon Gay Lussac was very strong. In 1809, on the death of Fourcroy, Gay Lussac was appointed to succeed him as professor of chemistry at the *École* and in the same year he became professor of physics at the *Sorbonne*, a post which he was to hold until 1832 when he resigned upon his appointment as professor of general chemistry at the *Muséum*. When Gay Lussac succeeded Fourcroy in 1809, Thenard returned to the *École* as a professor and at about this time the two men were also working together at Arcueil, following up Davy's electrochemical researches on the alkali metals. As a result of their experiments Gay Lussac and Thenard discovered the element boron.

L. J. Thenard, who was of peasant stock, arrived in Paris at the age of seventeen in 1794 and became laboratory assistant to Vauquelin. With his own natural ability and with help from Fourcroy he rose by 1804 to become Vauquelin's successor at the *École*. He also held chairs of chemistry in the *Collège de France* and in the Science Faculty at the *Sorbonne*. It was often necessary then for scientists to hold several posts at the same time in order to make a satisfactory living! Much of Thenard's chemical work was on mineral and inorganic compounds. In 1818 he discovered hydrogen peroxide and later he analysed hydrogen persulphide. In the organic field he worked on the tartrates and on Cadet's arsenical liquid (an impure cacodyl compound obtained by distilling arsenious oxide with potassium acetate). This compound had been described by Cadet in 1760 and it was to be fully investigated later (in 1837) by Bunsen, in connection with the isolation of the cacodyl radical.

The acids obtained by distilling fats were also investigated by Thenard, who found that impure acetic acid was commonly formed as one product of this process. A different acid was obtained however, when tallow was distilled and this he called sebacic acid. Berzelius wrongly suggested that this was benzoic acid contaminated by some unknown but separable substance derived from the fat. The true identity of sebacic acid as a di-carboxylic acid, was confirmed by Dumas and Peligot who gave the formula ($C_8H_{16}(COOH)_2$) in 1834. In a study of biliary calculi and the bile itself Thenard discovered a resinous substance and another peculiar compound with a bitter-sweet taste which he called *picromel*. This substance, which was also recognised by Berzelius and by Tiedmann and Gmelin in their essay on digestion published in 1826, was probably a mixture containing sodium taurocholate. Taurine ($NH_2.(CH_2)_2.$ SO_3H), was found in bile by Tiedmann and Gmelin. In 1823 J. B. A. Dumas became lecture assistant to Thenard at the *École*. Dumas was to become the leading French organic and animal chemist of the nineteenth century, with a reputation in this field surpassed only by those of Berzelius and Liebig. In organic chemistry Dumas proposed the theory of substitution in opposition to the radical theory which was based on Berzelius' dualistic chemistry. German chemists, including Liebig, on repeating Dumas' experiments could only confirm his results and by 1867 Liebig had accepted Dumas' concept of substitution. In 1835 Dumas succeeded Thenard as professor at the *École* and at the same time he was appointed professor of organic chemistry at the *École de Médecine*. He provided his students with systematic practical work at the *École* from 1832, before he became professor there and later at his own expense in his private laboratory in Rue Cuvier.

The relationships between plants and animals

J. B. Boussingault became professor of chemistry at the Conservatoire des Arts et Métiers in 1839 and he then began to collaborate with Dumas in researches on the composition of the air and on a variety of chemical and agricultural problems. The interest of these two chemists in the air was concerned with its role as the medium of support for the life of both animals and plants and when Dumas gave his final lecture in the course at the *École de Médecine* in 1841, he and Boussingault published a joint essay on the comparative physiology of animals and plants and their relationships to each other and to the atmosphere.

TABLE VI

Animals	*Plants*
Combustion apparatus.	Reduction apparatus.
Consume oxygen, neutral nitrogen compounds, fats, sugar, gums and starch.	*Produce* oxygen, neutral nitrogen compounds, oils, sugar, starch and gum.
Produce heat and electricity.	*Absorb* heat and electricity.
Change organic matter to mineral substances.	Change mineral matter to organic substances.
Give up elements *to* the air and soil.	*Take up* elements *from* air and soil.
Exhale carbon dioxide, water etc.	Fix carbon dioxide, water etc.

When this essay was published Liebig accused Dumas of plagiarism because much of the work which it covered had already been described in his *Organic Chemistry in its relations with Agriculture and Physiology*, published one year earlier in 1840. This was in fact an unjust accusation and the French chemists were able to show that their essay summarised the work which they had been doing both separately and in collaboration for some years. Most of their ideas had been published even before 1839.

Dumas and Boussingault had been especially interested in determining the nutritive value of feeding stuffs with particular regard to farm animals. This involved the difficult problem of the origin of the nitrogen of animal tissues, especially in herbivores such as the cow, sheep and horse. There seemed to be three main ways in which nitrogen might find its way into the tissues, viz., from the food, from the air during respiration, or by the transformation of other elements in the body of the animal into nitrogen under the influence of the vital force. Many experiments had already been made to try to settle this question. In some the composition of the food was compared with that of all the excreta in order to determine the intake of nitrogen. In others inspired and expired air were compared with respect to their relative nitrogen contents.

The origins of nitrogen in the tissues

About 1832 I. F. Macaire and François Marcet, son of Alexander Marcet the physician, examined the problem, beginning with the nitrogen content of vegetable and animal foods. They came to the conclusion that the distinction was one of degree, for whereas animal foods contained a high proportion of nitrogen, vegetable foods contained only a little. They found that the digestive process gave rise to forms of chyle which differed in appearance according to the nature of the food, but the differences were only superficial, for when the chemical composition of the chyle of a horse eating only grass was compared with that of a dog fed entirely on animal matter it was found that the two were almost identical. It seemed that whatever the nature of the food taken in, the body formed chyle of about the same elementary composition from it.

TABLE VII

	Chyle of dog (animal food)	Chyle of horse (grass only)
C............	55.2	55.0
H............	6.6	6.7
O............	25.9	26.8
N............ (azote)	11.0	11.0

The process took much longer and required much larger quantities of food for the herbivore than for the carnivore and there were considerable differences in the excreta. It was found that the carnivore yielded about five times as much nitrogen as the herbivore, thus:

TABLE VIII

	Dog	Horse
C............	41.9	38.6
H............	5.9	6.6
O............	28.0	29.0
N......(azote).	4.2	0.8
Mineral matter	20.0	25.0

From these results it appeared that the herbivore must be able to treat its food so that almost all the nitrogen in it was absorbed whilst the carnivore could afford to throw out of its body a higher proportion of the nitrogen contained in its food. As to whether animals might not obtain at least part of their nitrogen from the air Macaire and Marcet attempted to derive evidence from a comparison of the elementary composition of chyle and blood into which it was converted during respiration. Having made elementary analyses of the blood of sheep, rabbits, horses, cows and dogs, they concluded that the composition of blood was the same in all cases within the limits of experimental error.

TABLE IX

	Arterial Blood	Venous Blood
C............	50.2	55.7
H............	6.6	6.4
O............	26.3	21.7
N......(azote).	16.3	16.2

Elementary composition of the blood of the rabbit

These figures compare closely with those for chyle above; the proportion of carbon in chyle and in venous blood is very similar and consequently it might well be that the action of respiration is the same on these two substances, changing each into arterial blood by reduction of the carbon content. Turning to the nitrogen it is seen that there is more of this element in blood than in chyle and it might therefore be reasonable to assume that respiration causes a rise in the proportion of nitrogen. Since a very small proportion of chyle was mixing with the blood at any one time the absorption of nitrogen from the atmosphere had escaped notice, but Macaire and Marcet concluded that it occurred nevertheless.

Some physiologists had thought that animals might be capable of creating nitrogen from other elements in their food. This idea was no doubt supported in part by the uncertainty which then existed as to whether or not 'azote' was an element. The same doubts had also been entertained with respect to other elements. Thus, Vauquelin had fed fowls on grain alone and on weighing the eggshells he had found that a great deal more calcium carbonate was given out by the fowl than was contained in its food. This led him to speculate that silica might change into chalk during the process of digestion in the fowl. If this were the case with respect to the production of azote in the chyle it would follow that the nature of the food taken in by the animal would be a matter of indifference and all animals would be nourished equally well by any food whatever. As Macaire and Marcet pointed out, this had already been shown to be false. In some experiments which the physiologist Magendie had carried out in 1816, dogs had been fed on substances such as sugar, olive oil and butter which do not contain nitrogen and in every case the animals had wasted away and died after about thirty-six days. The dogs were found to be producing chyle but this was nevertheless incapable of supporting life. This seemed to indicate that most of the nitrogen required by the animal was derived from its food and when this element was absent the animal was incapable of supplying the lack. Since Magendie's experiments had been confined to dogs, Macaire and Marcet decided to test the effects of feeding a herbivorous animal on food entirely devoid of nitrogen. They chose a yearling lamb which they fed on bread, sugar and gum; after eleven days it began to show signs of malnutrition and on the twentieth day of the experiment it died. From such investigations it became clear that neither carnivores nor herbivores were capable of sustaining life on a diet devoid of nitrogen, although it was quite possible that a small proportion of nitrogen was absorbed during respiration.

Nutrition of farm animals

Dumas and Boussingault took up the problem of the nutritive value of feeding stuffs with respect to their nitrogen content in the case of farm animals. Boussingault first attempted to determine the proportions of nitrogen in different animal feeds so as to fix a comparison between them and so help farmers to substitute a cheaper kind of forage for an equivalent quantity of a more expensive one. All vegetable foods contain some nitrogen and Boussingault assigned to each type of forage a number by which he indicated its relative nutritive value. He ignored the non-nitrogenous parts of the plants and insisted that real nutritive value could only be measured in terms of nitrogen.

The next step was to feed a cow for a month on potatoes and grain, analysing all the food supplied and *all* the products from the cow. By comparing the nitrogen contents of the food with that of the products Boussingault came to the conclusion that the food had supplied nearly all the nitrogen required by the cow and that little, if any, was absorbed from the atmosphere. The same conclusion was reached in experiments on a horse and a few years later in some tests on a turtle. Boussingault showed that when a cow was fed on a nitrogen deficient diet it steadily lost weight because it was forming milk at the expense of its tissues. He decided that it would not be possible to replace the more expensive fodder such as hay, oil-cake and grain by cheaper materials like roots and potatoes.

In another series of experiments carried out in 1846, Boussingault used ducks. The birds were first kept without food for thirty-six hours and were then fed to repletion with known weights of specific foods. They were then placed in boxes in which their droppings could be collected and weighed. Boussingault thought that the difference in weight between the food and the excreta would indicate the amount of matter absorbed by the bird or eliminated in respiration. The ducks were then killed at different intervals after the beginning of the experiment and their internal organs were examined. From these experiments it became clear that nitrogenous substances such as albumen, fibrin and casein, although necessary to the diet, were not complete foods in themselves and must be accompanied by other foods like starch, organic acids and gelatin. The latter had for long been regarded favourably as an economical means of extending the diet of the poor. D'Arcet, who had devised an improved method of extracting this substance from bones, energetically advocated its use as a food, but there was doubt as to its real nutritional value. A Commission was therefore set up under the leadership of Magendie, to look into the true merits of gelatin as a food. This Commission reported in 1841 its conclusions that gelatin, albumen or fibrin taken singly were capable of supporting animal life for short periods only; the other constituents of flesh such as fats and salts were also essential for a balanced diet. Bones contained all the requirements of such a diet but it seemed that all the known methods of preparing food from bones reduced their nutritive value and hence the attempt to supplement the diet of the poor from this source was not likely to be very successful.

Chevreul and the chemistry of fats

The chemical constitution of the fats was worked out in a model series of experiments by M. E. Chevreul, who had been a pupil of Vauquelin at the *Muséum* from 1803. Chevreul spent his long working life at this institution, for in 1810 he became an assistant there, professor in 1830 and finally Director in 1864. He also had charge of the Gobelins dye factory where he worked on the properties of dyes and colours. Vauquelin had introduced him to the study of natural colouring matters, but in 1811 Chevreul turned his attention to the fats. Over the next twelve years he produced a series of researches concerned with these animal products which was so logically constructed and so complete that Berzelius called it a '... model for young chemists, the best and most completely executed that chemistry can show.'

The fact that an acid was always produced during the destructive distillation of a fat seemed to indicate that these substances contained an acid principle, but Fourcroy had suggested that the great heat of the fire caused an alteration in the fat itself. It was also known however, that fats reacted with alkalies to form soaps, compounds which Berthollet regarded as analogous with salts. This had led to the suggestion that the *whole fat* behaved as an acid, yet the constitution of the fats seemed to be more complex than this, for Scheele had observed that a sweet-tasting principle (glycerol) was produced when fats were heated with litharge.

Chevreul began with a potassium soap he had obtained from pig-fat by treating it with caustic potash. He found that part of this soap dissolved in water and part was deposited as white pearly crystals which he was able to decompose by adding acid. From this a solid fatty acid, soluble in alcohol, was separated without the aid of heat which was known to cause decomposition. Chevreul first called this acid margarin (a name later adopted for a now familiar product), but he later changed the name to margaric acid. From that part of the soap which had dissolved in water he went on to isolate another fatty acid to which he gave the name oleic acid. Thus he was able to show that the original soap contained at least two products formed by the combination of potash with two different acids. Since the fat was changed from a neutral substance into a fatty acid salt and the sweet principle under the influence of potash, it seemed to Chevreul that the fats were composed of fatty acids combined with the sweet principle. He went on to determine the relative weights of fats which a given weight of an alkali such as potash, soda or lime could saponify and concluded that the fatty acids were typical acids capable of yielding salts with the bases. The chemistry of the fats appeared to show close analogies with the well-known reactions of inorganic compounds.

When treated with hot alcohol, pig-fat was found to separate into two fractions, of which one was a crystalline solid and the other was an oil. This had also been noted by Braconnot, Director of the Botanic Garden at Nancy, who called the solid part of the fat 'absolute tallow' and the liquid part 'absolute oil'. In order to separate these components Braconnot had used a rudimentary form of chromatography, imbibing the oily parts of various fats, oils and butters onto blotting paper and thus separating them from the tallow. He found that the difference between the various fats could be expressed in terms of the relative proportions of their solid and oily constituents. This work, which was published in 1815, displeased Chevreul who claimed priority for his own researches on the same subject. Chevreul examined the fats of sheep, beef, goose and jaguar, as well as human fat and in each case he found that they could be separated into solid and liquid fractions. The differences between the fats were due to the different proportions of these constituents and it was on this basis that the wide variety of oils, fats and butters must be explained. On saponification all fats yielded the sweet principle and soaps formed from the two fatty acids, margaric and oleic. Chevreul later discovered several other acids in dolphin oil and in cow's and goat's butter; he gave the name 'glycerin' to the sweet principle and found that it could be isolated from the water in which it was dissolved after the soap had been separated.

Chevreul also investigated the three forms of fat which had been regarded as varieties of 'adipocire'. These were the crystalline substances found in biliary calculi, spermaceti and the fat formed in corpses. He found that the first of these could not be saponified and was therefore not to be thought of as a fat at all. Spermaceti could be saponified, but did not yield the sweet principle and was consequently not to be regarded as a fat, but the fat in corpses behaved as a mixture of fats in the usual way. Thus the three substances which Fourcroy had thought to be varieties of the same compound were shown by Chevreul to be distinct materials.

Since he had now identified a number of different fatty substances, Chevreul attempted to supply a nomenclature for them. Thus, he proposed to call the crystalline substance found in biliary calculi, *cholesterin*, whilst spermaceti became *cetin*; the solid fat he called *stearin* and the liquid one *elain*. The acid derived from the solid fat was to be called *margaric acid* and that obtained from the liquid fat was named *oleic acid*. The salts and esters of these acids were to be called *margrates*, *cetates* and *oleates* in recognition of their similarities to inorganic compounds.

Until 1820 Chevreul was concerned only with the separation of 'immediate principles' such as the fatty acids from the fats and it was not until later that he used elementary analysis. Since the fatty acids were so similar in their chemical properties their separation required considerable skill. Chevreul used solvents to separate his immediate principles and he characterised them by means of their melting-points. The use of a physical constant as a criterion of purity and a means of identification was an important step in the development of organic analysis and Chevreul was a pioneer in the use of the melting-point for this purpose. From his elementary analyses he discovered that the products of saponification contained more hydrogen and oxygen in the proportions of water than the original fat and from this he concluded that the elements of water were taken up during the process of saponification. In 1823 Chevreul summarised his results in his *Recherches Chimiques sur les corps gras d'origine Animale* in which he drew a number of conclusions. For example, he pointed out that the fatty acids showed a deficiency of oxygen and an excess of hydrogen and that this was clearly a blow to Lavoisier's oxygen theory of acidity. The classification of organic substances into oily, neutral and acidic was seen to be premature and the chemical nature of a substance could not be determined merely from the relative proportions of its elements. Chevreul compared the fats with esters, suggesting that glycerol must be analogous with alcohol. During saponification the acidic and alcoholic parts of the ester were separated and the soap formed was a true salt of the fatty acid.

Chevreul also suggested a method of classification for organic substances which resembled the Linnaean system used in biology. Compounds were to be divided into species, varieties and genera. A species was to include substances identical in their chemical nature and the proportions and arrangements of their elements; varieties were considered to occur within species when there were several crystalline forms or other minor differences in the physical properties of the members of the species, whilst genera were to include collections of species with one or more common property. Chevreul suggested that his classification should be made the basis of a set of characteristics which were to include,

(a) elementary composition,
(b) physical properties,
(c) chemical properties, types of reaction undergone, stability to acids, alkalies, heat, etc.,
(d) physiological properties.

In this way he hoped to establish a method of classification which would be most suitable for those compounds with the closest ties with living things.

By his researches Chevreul organised the study of the animal fats and showed that these natural substances in some respects resembled mineral and inorganic compounds. The fats were now seen to be a class of closely related compounds containing chemical groups which appeared repeatedly in similar structures. This observation lent support to the radical theory of organic chemistry and it also upheld the view that there could be no fundamental difference between the chemistry of inorganic and organic or even organised substances. It now seemed quite feasible that new fats might be prepared synthetically according to the same pattern of combination. Such substances would then be represented by definite chemical formulae and would be capable of preparation in pure forms which could be identified precisely, for example by their melting points. This degree of precision in classifying organised matter had never before been possible, but it made clear the fact that the fats could be treated simply as chemical entities without the need to introduce the confusing issues of vitalism, even though they were substances which had been formed in the living animal body.

Chapter 7

Justus von Liebig

T H E C H E M I S T R Y of living things was given a new dimension in the work of Liebig and his school at Giessen. All the relatively isolated attempts to understand such processes and substances as were known to occur in living organisms were brought into relationships with each other and given quantitative expression by means of Liebig's famous 'equations'. Based firmly upon analytical data, Liebig's physiological chemistry went far beyond the immediate observations to suggest a theoretical and conceptual framework within which life-functions could be explained. Critics of Liebig's work asserted that it contained too much speculation, but there was undoubtedly a real need for some such conceptual scheme within which known facts could be fitted and from the use of which new ones could be predicted. In a very real sense Liebig's work can be regarded as a 'watershed' between the older, largely empirical animal and vegetable chemistry and the newly developing science of biochemistry. Liebig's work was by no means infallible or correct in every detail, but it served to point the physiological chemist in the right general direction. His importance both to organic and to physiological chemistry far outweighed that of any other nineteenth century chemist, so that he ultimately displaced and succeeded Berzelius as the great master of these subjects.

Liebig: organic chemist

Justus von Liebig was born at Darmstadt on 12th May, 1803. His father was a dealer in drugs and dyes some of which he prepared himself and he was therefore able to teach his son the art of making simple experiments in chemistry. At school

Liebig was unsuccessful and made little headway because he could not easily remember those things which he had only heard. Instead his mind turned to practical work and he found it possible to learn only that which he had worked out for himself as a result of his own observations. When he declared that he would be a chemist he was ridiculed for chemistry was not then a subject which could form the basis of a career in Germany. At best it was an ancillary to medicine, but Liebig, who had read widely if at random, believed that chemistry was worthy of study in its own right.

At the age of sixteen he entered the University of Bonn where he began to study under C. W. G. Kastner. When Kastner left Bonn for Erlangen in 1820, Liebig followed him but soon became dissatisfied with his lack of progress in chemistry, despite the fact that he had tried to gain as much knowledge as possible in this subject. In 1822 Liebig took his doctorate in philosophy at Erlangen and also about this time he published his first chemical paper on fulminate of mercury. It was this work which opened the way for him to study abroad for it was brought to the attention of the Grand Duke of Hesse-Darmstadt who provided Liebig with money to enable him to pursue his studies. Liebig had come to the conclusion that it was impossible to become a chemist in Germany for science in that country was under the influence of the speculations of *Naturphilosophie*. Thus it was that Liebig went to Paris where there was, as we have seen, a group of experimentalists who were rapidly developing the sciences of physics and chemistry. There were at that time no laboratories devoted to the instruction of students and the only means of obtaining entry to a laboratory was by introduction to one of a small number of private establishments or as a protégé of one of the famous chemists of the day. It was on the introduction of Thenard that Liebig was enabled to enter the private laboratory of Gaultier de Claubry, (formerly belonging to Vauquelin). Here he continued his experiments on explosives and soon published another paper on the subject. Then in the summer of 1823 he met Humboldt and from that time onwards his course was greatly advanced. He was introduced to the laboratory of Gay Lussac at the Arsenal and it was whilst working with this great master that Liebig conceived the idea that he would like to set up a school of experimental chemistry in Germany. In the following year, 1824, Humboldt recommended him as professor of chemistry in the small University of Giessen and here he set out to realise his dream. Liebig's laboratory at Giessen was to become famous, both for his teaching and for the vast number of organic analyses which poured from it during the next twenty-eight years until 1852 when he moved to Munich. This was the first German chemical laboratory in which systematic practical instruction was given.

At first Liebig had no private laboratory for his own work and was forced to carry out his researches in the same room as his pupils. There was no heat in the balance room and all the accommodation was too small. In the lecture room the students at the front had their ink-pots on the lecture bench whilst in hot weather those at the back could drop out of the windows and go to an adjacent beer cellar for some refreshment. There was no fume chamber in the laboratory and experiments with poisonous and noxious vapours had to be conducted in the open air. In the laboratory the workers wore hats to protect their hair from flying sparks given off by the fanned charcoal furnaces. Nevertheless the laboratory was always a hive of industry and there were so many students that they were able to form their own societies and had little contact with the other students of the University. Many of Liebig's students went on to become famous in their own right and some (e.g. A. W. Hofmann at the Royal College of Chemistry, established in London, 1845) proceeded to set up schools of chemistry along similar lines to those of Liebig himself.

We have already described Liebig's method of organic

analysis which was developed from earlier procedures and depended upon the total oxidation of the carbon and hydrogen of the organic matter, followed by the quantitative estimation of the carbon dioxide and water vapour formed. This then led to an empirical formula for the organic compound, although there were difficulties due to unreliable atomic weight data. For a long time Liebig's interests were in pure organic chemistry. In 1832 he worked with Friedrich Wöhler, his life-long friend, on the chemistry of oil of bitter almonds, discovering in the process the radical benzoyl (C_6H_5CO) and the first known glucoside, amygdalin. Also in collaboration with Wöhler about 1838 he elucidated the chemistry of uric acid derivatives and so initiated the study of the purines.

Liebig and Wöhler on the derivatives of uric acid

The red and purple compounds formed when uric acid reacts with nitric acid were noticed by Scheele and had been used as identification tests for uric acid since his time, but the chemical compositions of these compounds were unknown until Liebig and Wöhler worked on them in 1837-8. Before this, Prout had studied them with limited success and in 1818 Gaspard Brugnatelli had isolated a colourless acid from uric acid whose salts were all red or purple in colour. The acid was named 'erythric acid' and Prout claimed that it was different from a colourless acid which *he* had isolated. This acid also gave red or purple salts and Prout named it 'purpuric acid', suggesting that Brugnatelli's acid was probably some compound between purpuric acid and nitric acid from the fact that oxidation of purpuric acid with nitric acid produced a substance identical in properties to erythric acid.

In their investigations of 1837-8, Liebig and Wöhler identified and studied no fewer than thirteen related substances, all derivatives of uric acid. Chemical analyses of all these compounds were made, their empirical formulae were worked out and the chemical relationships between them were expressed in terms of 'equations' similar to those which Liebig was to use four years later in his *Animal Chemistry*. The ammonium salt of Prout's purpuric acid was called murexide by Liebig and Wöhler and formulated $C_{12}N_{10}H_{12}O_8(C=6)$. It was clearly distinguished from erythric acid which the German chemists renamed alloxane $(C_8N_4H_8O_{10})$. Prout's purpuric acid, now called murexane $(C_6N_4H_8O_5)$, could be obtained from murexide by acid hydrolysis in a reaction which Liebig and Wöhler represented thus:

TABLE X

2 atoms murexide $C_{24}N_{20}H_{24}O_{16}$	1 atom	alloxane	$C_8N_4H_8O_{10}$
	1 atom	alloxantin	$C_8N_4H_{10}O_{10}$
	1 atom	murexane	$C_6N_4H_8O_2$
11 atoms water	$H_{22}O_{11}$	1 atom	urea	$C_2N_4H_8O_2$
	2 atoms	ammonia	N_4H_{12}

| $C_{24}N_{20}H_{46}O_{27}$ | $C_{24}N_{20}H_{46}O_{27}$ |

Liebig and Wöhler suggested that many of the derivatives of uric acid could be inter-related through a radical which they called 'uril' $(C_8N_4O_4)$. By the addition of simple molecules such as water, ammonia, oxygen and so on, to this radical it was possible to obtain the empirical formulae of many of the uric acid derivatives. The importance of this observation for animal chemistry lay in the fact that though they were obtained from animal sources and must therefore be regarded as organised matter, these compounds clearly obeyed the same kind of laws of combination as were known to hold amongst mineral and ordinary organic substances. The formulae of these compounds were all very similar and the changes relating them were numerous and intricate, nevertheless the subject formed a direct link between organic chemistry and the functions of life.

Liebig turns to agricultural and physiological chemistry

Liebig's work at Giessen formed the foundation upon which he based his later investigations into physiological chemistry. One of the distinctive features of his method was the attempt to make the subject more quantitative by the use of equations in which the empirical formulae of compounds were added and subtracted. Such equations did not, of course, explain the minutiae of the physiological processes but they placed the overall changes occurring in animal metabolism on a quantitative basis.

Early in 1840 Liebig declared himself tired of laboratory work and said he intended to study the applications of organic chemistry—a subject more appropriate to the later periods of life. He found the chemical investigation of physiological functions more fruitful than the study of innumerable substitution products which never appeared in nature. In the same year Liebig published his first important work on agricultural chemistry and outlined at the British Association Meetings his ideas on this subject. He wanted to make agriculture more scientific and less empirical. At this time it was commonly believed that plants derived their nutriment from humus (vegetable mould) and the fertility of the soil was generally ascribed to the quality of its humus content. In England however, the function of humus as the main plant nutrient was less widely accepted, since Priestley had long before shown that plants derive much of their carbon content from the atmosphere.

Liebig showed that humus is of very variable composition and that it would only dissolve in water when freshly precipitated. After drying and on exposure to freezing temperatures, humus became insoluble in water. Ordinary climatic conditions would be enough to bring about this change and in fact it was observed that very little organic matter would dissolve in water which was allowed to percolate through a good fertile soil. Since plants were known to take up their nutriment in dissolved form, it was clear that humus could not in general be the main source of plant food. Liebig therefore set out to determine what did provide the source of plant nourishment in the soil and he concluded that this was to be found in the mineral salts remaining after the decomposition of the humus itself. Plants were found to need, besides carbon dioxide and water vapour from the air, inorganic salts including phosphates, nitrates, ammonium salts and alkalies such as lime, magnesia and potash. He showed that organic manures could be replaced by the right combinations of these mineral compounds because these are the necessary substances for plant growth and organic matter is slowly decomposed into such mineral compounds by the action of atmospheric oxygen and bacteria. The results were obtained from controlled experiments on the farm using mineral manures manufactured in Liebig's own works. Their application in practice led to considerable improvements in crop yields.

'Animal Chemistry', 1842

Using experimental results, calculations and deductions, Liebig worked out relationships between plants, the soil and the atmosphere. Thus, plants take up mineral substances from air and soil, from these they fabricate starch, sugar and gum, at the same time releasing oxygen to the air. Animals, on the other hand, need prefabricated food either from the plants or from the flesh of other animals. As we have seen, Dumas had reached a similar conclusion about the same time.

Whilst he was working on the chemistry of agriculture Liebig also studied the applications of chemistry to the physiology and pathology of animals. By using the accumulated results of organic analysis he was able to make a more quantitative study of the subject which led him to a better understanding of the relationships between plants, animals and the environment. At every point Liebig tried to simplify and generalise his results as far as possible. In 1842, two years after his first book in this field, Liebig published his most famous and controversial work, *Animal Chemistry, or Organic Chemistry in its applications to Physiology and Pathology*. Like his earlier work on agricultural chemistry this book was soon translated into English and it rapidly became well-known both in this country and abroad. But Liebig had made his *Animal Chemistry* so comprehensive that it was seen to challenge most of the orthodox theories of physiology and because of this the work soon became the centre of controversy. Compared with other works on the same topics this book was much more quantitative. It showed precisely what the food of animals contained and the chemical composition of their tissues and excretions. It proved by easy stages how the food could be changed into animal tissues with the release of energy and how the tissues themselves were converted into waste products and energy. The whole argument of the book was carefully documented with chemical equations derived from the results of analysis, both by Liebig and others until it seemed that animal chemistry could be expressed in terms of simple chemical equations. These equations were regarded as the most novel feature of the book, yet after only a few years it appeared to contain far too much speculation and the equations were abandoned.

In the *Animal Chemistry* Liebig stated that the oxidation of carbon to carbon dioxide and of hydrogen to water was *the source of animal heat*. This idea was not new of course, since it had already been stated and investigated experimentally by Lavoisier and Laplace in 1780. However, more exact measurements since that time had indicated that the intake of oxygen during respiration was insufficient to produce all the heat evolved by the animal. Nevertheless, Liebig boldly stated the principle in capital letters as though there could be no doubt about it. His basis for doing so was an unexpressed conviction about the conservation of energy in the living body. Like Dumas, Liebig divided foods into two main categories distinguished by the presence or otherwise of nitrogen. Albumen fibrin and casein which contain nitrogen were regarded as *plastic* elements of the food and were thought to provide the material for the growth and repair of tissues, whilst substances such as fats, sugar, starch and gum, which do not contain nitrogen, were called *respiratory* foods and were thought to provide the body with its energy when they were oxidised. Respiratory foods must lead to the evolution of carbon dioxide and water vapour during respiration. In the body however, there was also the degradation of tissue material by oxidation—a process which Liebig called the 'metamorphosis of tissue'. This process gave rise to the release of heat energy and the production of excretory products including uric acid, urea, oxalates and so on. These appeared mainly in the urine. By analysing the tissues and excretory products, Liebig claimed to be able to account for the metabolic processes and express them by means of his equations. He showed great ingenuity in selecting just those processes which fitted in well with his theory and many of his equations seemed to express the overall metabolic changes with a remarkable degree of precision.

In 1837-8 G. J. Mulder the Dutch animal chemist showed that the elementary compositions of albumen, fibrin and casein, whether from animal or vegetable sources, were almost identical. These three substances contained the same proportions of carbon, hydrogen, oxygen and nitrogen, but differed in their content of sulphur and phosphorus. This discovery led Mulder to conclude that these substances, together with muscle tissue and blood, all contained a common radical which, following Berzelius's suggestion, he called 'proteine' and to which he gave the highest importance in the formulation of animal tissue. At first Liebig embraced this theory with enthusiasm. It seemed to offer a simple explanation of the manner in which plant food could serve as food for animals and Liebig formulated the proteine radical as $C_{48}N_6H_{36}O_{14}(C=6)$. Albumen and fibrin, important constituents of the blood, were thought to contain proteine combined with sulphur and phosphorus, whilst gelatin, hair and horn contained proteine united with ammonia. But, later on when all attempts to isolate the proteine radical had failed, Liebig denounced the idea as vigorously as he had at first defended it.

$$
\left.\begin{array}{l}
\text{5 atoms proteine} \\
\text{15 atoms starch} \\
\text{12 atoms water} \\
\text{5 atoms oxygen}
\end{array}\right\} = \left\{\begin{array}{l}
\text{9 atoms choleic acid} \\
\text{9 atoms urea} \\
\text{3 atoms ammonia} \\
\text{60 atoms carbonic acid}
\end{array}\right.
$$

In detail:

5 atoms proteine	$5(C_{48}N_6H_{36}O_{14}) = C_{240}N_{30}H_{180}O_{70}$
15 atoms starch	$15(C_{12}H_{10}O_{10}) = C_{180}H_{150}O_{150}$
12 atoms water	$(12(HO) = H_{12}O_{12}$
5 atoms oxygen	$= O_5$

$$C_{420}N_{30}H_{342}O_{237}$$

9 atoms choleic acid	$9(C_{38}NH_{33}O_{11}) = C_{342}N_9H_{297}O_{99}$
9 atoms urea	$9(C_2N_2H_4O_2) = C_{18}N_{18}H_{36}O_{18}$
3 atoms ammonia	$3(NH_3) = N_3H_9$
60 atoms carbonic acid	$60(CO_2) = C_{60}O_{120}$

$$C_{420}N_{30}H_{342}O_{237}$$

TABLE XI

Liebig and Animal Chemistry: the metamorphosis of food and tissues

Liebig adopted the formula $C_{48}NH_{39}O_{15}$ for flesh and the protein part of blood. He was able to show that the same composition could be obtained by adding together choleic acid $(C_{38}NH_{33}O_{11})$, the chief constituent of the bile and ammonium urate $(C_{10}N_5H_7O_{16})$, the two compounds which were known to be the principal products of the metamorphosis of muscular tissue. In the lower animals such as the amphibia and the insects, the process is arrested at this stage and ammonium urate is excreted in quantity, but in the higher animals the ammonium urate is further oxidized to urea by a series of intermediate steps. Liebig was not satisfied to state this in general terms, even though he could demonstrate it from his observations. He wanted to be able to account for the changes quantitatively in such detail that he could show that nothing was left over. Thus he claimed that when proteine and starch were oxidized together in the presence of water, they produced choleic acid, urea, ammonia and carbonic acid only, (Table XI).

By studying the empirical formulae of respiratory foods such as the fats, starch, sugar and gum, Liebig was able to draw a very interesting conclusion which provided an answer to at least two outstanding problems. These were the origin of fat in the bodies of the herbivores and the deficit of oxygen in the production of animal heat by respiration. He noticed that it was only when an animal received an excess of non-nitrogenous food that it began to form fat. The proportions of carbon and hydrogen in starch, sugar and fats were known to be very similar. Chevreul had found 79 per cent of carbon and between 1.1 per cent and 11.7 per cent of hydrogen in fats, whilst 79 parts by weight of carbon was united with 10.75 parts of hydrogen in starch and with 11.82 parts of hydrogen in gum and sugar. In the latter compounds there was a higher proportion of oxygen than in the fats and Liebig argued that merely by losing some of their oxygen, starch, gum and sugar could be transformed into fats. The separated oxygen could then be used by the animal to oxidise carbon and hydrogen in the tissues and so account for the discrepancy between the inspired oxygen and the quantity required for the production of animal heat. This neat piece of speculation illustrates Liebig's work; it appeared very plausible, but to establish it as a process taking place in animal metabolism would have been very difficult indeed. Many other parts of Liebig's book were susceptible to similar criticisms. In fact, the most important service which Liebig rendered to animal chemistry was to provide a body of theoretical ideas, founded upon chemical analysis but requiring a tremendous amount of further detailed investigation. It was not possible to accept Liebig's *Animal Chemistry* as a simple expression of the truth, but regarded as a fairly coherent set of reasoned hypotheses which could be submitted to experimental investigation, this book had an important part to play in the rise of biochemistry.

Reception of Liebig's 'Animal Chemistry'

Amongst the chemists of his day Liebig recognised only the authority of Berzelius as the great master of physiological chemistry and he hoped that the older chemist would in time willingly pass on the torch to him. During 1840-1 the two chemists were in frequent correspondence with each other and Liebig explained his new ideas on agricultural and physiological chemistry to Berzelius who showed genuine interest in them. However, the Swedish chemist thought there were doubts about the validity of some of Liebig's ideas which were based on an insecure experimental foundation. Liebig wished to dedicate the *Animal Chemistry* to Berzelius with a long and flattering tribute supporting him at a time when others were already beginning to attck his views. Berzelius accepted the dedication but declined the flattery on the grounds that it would prejudice any critical comment he might later wish to make. Each year, for more than twenty years it had been Berzelius' practice to summarise and comment upon the leading chemical publications of the preceding year in his annual *Jahresbericht* and his judgements in these summaries had long exerted a powerful influence on the development of chemistry. Liebig was naturally anxious to have a good review from Berzelius, especially since animal chemistry was the subject in which the latter had excelled. It must therefore have been a shock to him to read first Berzelius' comment about '... This facile kind of physiological chemistry ... created at the writing table ...' in the *Jahresbericht* for 1843 and strong criticism of his book in the following year.

Berzelius' main criticism was that Liebig had used all his considerable powers of persuasion to make hypotheses appear proven facts. Liebig's equations were premature, he said, because the data on which they were based were not yet accurate enough for such calculations to be made. Liebig took Berzelius' criticisms as a personal attack and in 1844 he replied to them in his own *Annalen der Chemie und Pharmacie*, adding some counter criticisms of Berzelius' work in animal chemistry. In this way a bitter quarrel began between the two leading organic chemists. Berzelius, who had laboured to establish animal chemistry by careful attention to the details of chemical composition and properties, regarded Liebig's approach as basically unsound because he was trying to erect a comprehensive theory from too few and uncertain experimental results. It is also quite possible that Berzelius resented the bid by Liebig to take over the lead in the field of chemistry in which he had for so long been the pioneer and in which he had become the master. This unfortunate quarrel serves to illustrate the fact that scientists too are sometimes motivated by the baser human qualities.

There were few other chemists or physiologists to whom Liebig's *Animal Chemistry* meant so much on a personal level and many accepted it on Liebig's authority as an organic chemist. Friedrich Mohr, chemist and pharmacist, friend of Liebig and one of his former pupils, accepted the new theories of animal chemistry at once and began to teach them to his medical students as the basis of a new era of rational medicine. Later, when two members of his family fell ill, with symptoms of emaciation and diarrhoea, Mohr deduced from Liebig's theory that the emaciation came from the failure to reabsorb the bile compounds with consequent extra degradation and loss of muscle tissue during respiration. When the treatment which he prescribed based upon this idea proved successful, Mohr wrote enthusiastically to Liebig in praise of his theories ... 'Everything fits, your theory will never be overturned, it is *true*.'

Lyon Playfair, who had also been one of Liebig's students at Giessen and had later attended him on his visits to England in 1837 and 1842, read an extract from the *Animal Chemistry* to the British Association in 1842, commenting that it was full of interesting and important views and that even the equations were the results of careful calculations and not merely the product of a brilliant imagination.

Application of Liebig's Physiology to Medicine

Henry Bence Jones, a London physician, had studied in Liebig's laboratory at Giessen for a brief period in 1841. He had been greatly impressed by the new theories of animal chemistry which Liebig was then developing, and became convinced that this was the right approach to the chemistry of life. 'My first conversation with Professor Liebig on his views on physiology had filled me with admiration and appeared like a new light where all had been confusion and incomprehensible before,' said Bence Jones. The concept of the oxidative metamorphosis of tissues seemed to provide the fundamental mechanism for all the functions of life. So inspired was he with the apparent power of this concept that upon his return to England he rapidly produced his first book *On Gravel, Calculus and Gout: chiefly an application of prof. Liebig's Physiology to the Prevention and Cure of these Diseases*. Bence Jones developed Liebig's ideas as an aid to the medical diagnosis and treatment of a group of diseases which were a serious scourge at that time.

Starting with Liebig's suggestion that uric acid was formed by oxidation of muscular tissues at times when the vitality was low, Bence Jones argued that it should be possible to correct this by tonic medicines which would augment the vital powers and at the same time promote the oxidation of uric acid to urea. This could be encouraged by the use of nitrous oxide solution and iron salts, whilst the availability of oxygen could be increased by reducing the quantities of fats, starch and sugar consumed. He was offering these ideas as a contribution to the treatment of urinary diseases based on Liebig's physiology.

Criticism from the Physiologists

Liebig's persuasive style and the apparent precision and simplicity of his equations won support from some physiologists and similar equations began to appear in textbooks of physiology. On the other hand there were those who were sceptical of Liebig's chemistry. Few doubted that chemical reactions played a part in life-functions, but they were not life itself—the old tradition of vital force could not be brushed away so lightly. Liebig had himself accepted the need for the vital force although he viewed it as another physical force comparable with gravity, electrostatic force or magnetism. Physiological problems were generally thought to involve a vitality which lay not in elementary chemical compositions but in the morphology of tissues and organs. Many physiologists thought that whilst Liebig's work was certainly about chemistry it was not in any sense about physiology. As always, the most valuable criticism of Liebig's work came from those who tried to make an objective assessment of it, neither accepting it unreservedly nor condemning it out of hand.

Amongst these the most influential was Johannes Müller, the great German master of physiology. Müller recognised that physiology was dependent upon other sciences and he held up the publication of the fourth edition of his famous handbook of physiology so that he could incorporate some of Liebig's ideas into it. He was particularly impressed by the new light which the *Animal Chemistry* threw upon the relationship between nutrition and respiration. This led him to emphasise the proportionalities between quantities of food, animal heat, motion and the amount of carbon used up in respiration. He incorporated some other work into his discussion in addition to that of Liebig. This included the earlier work of Prout and of Allen and Pepys on the relationships between respiration and exertion. Müller did not altogether accept the view expressed by Liebig that sugar not involved in the production of animal heat or energy is stored in the body entirely as fat—he did not discount the possibility that it might also be sometimes converted into lactic acid for example. When he realised that Liebig's work would entirely transfer the basis of physiology from comparative anatomy to physics and chemistry, Müller suddenly withdrew from physiology altogether and refused to publish any further editions of his handbook because he was unwilling to make such a drastic realignment. He devoted the remaining eighteen years of his life to anatomy.

Theodor Bischoff, a pupil of Müller recognised that Liebig's work contained much that was probable, although the speculative manner in which it was expressed did not amount to proof in the commonly accepted scientific sense. Unlike Müller however, Bischoff was prepared to change the basis of his physiological work and he introduced many of Liebig's ideas into his studies of this subject. In fact he was to become a pioneer in the application of the new ideas to physiological experimentation.

One of the fairest critics of Liebig's work was Otto Kohlrausch, who could see the significance of the *Animal Chemistry* as clearly as he could see the dangers of accepting its unsupported conclusions without question. He pointed out in 1844 that the greatest benefits would be derived by those who followed Liebig's example without slavishly copying his results. Kohlrausch showed that Liebig's accounts of experiments on animal heat were unsatisfactory and he pointed out that there was not enough information about the flow of the bile to claim it as a fact that reabsorption occurred. Liebig's ingenious use of empirical formulae was also challenged by Kohlrausch, who showed that it would be quite possible to produce an entirely different result applying the same techniques. So long as it was merely a case of proving the point by the addition of formulae, it would be possible to come to *any* desired conclusion. Kohlrausch doubted whether the processes of nutrition and respiration were in fact so closely linked together in the body as Liebig claimed since it was possible to vary the diet widely and yet the body was able to compensate for the variations. Nevertheless, Kohlrausch still believed that Liebig had done a great service for physiology by demonstrating that quantitative relationships between physiological processes undoubtedly exist.

Such criticism was important because it made even Liebig think again about some of his less well founded assertions, though he was not given to acknowledging the validity of doubts and criticisms expressed by his opponents. In 1846 he published the first part of the third edition of his *Animal Chemistry*. This was to be an enlarged edition, though it was never completed and it is worth noting that the equations which had played such an important part in the first edition were entirely absent in the third. The positive assertion that the bile is reabsorbed to provide material for respiratory processes was replaced in the new edition by experimental evidence on the absorption of soda and other soluble salts in the intestines. Liebig also placed less stress on the distinction between the nutrition of the herbivores and that of the carnivores; he corrected mistakes which had been due to his inadequate knowledge of physiology and in general qualified many statements which had been made too boldly in his first edition.

Problems in the theory of nutrition

There was one fundamental difficulty which confronted Liebig's theories of nutrition by 1846 which was not mentioned in the third edition of the *Animal Chemistry*. The proteine theory which had largely inspired his work nine years earlier was now beginning to prove untenable. Liebig and his students had been quite unable to isolate the proteine radical free from sulphur and phosphorus, even using Mulder's own methods and in several short articles in 1846 Liebig cast doubts on the real existence of this radical. This resulted in an acrimonious dispute between Liebig and Mulder who saw the basis of his methods of protein analysis threatened. In the meantime Laskowski, one of Liebig's pupils who had made a thorough study of Mulder's work, published a long paper in which he completely repudiated the proteine theory. Liebig now claimed that Mulder's theory that albuminous animal and plant materials are all identical in chemical composition had deluded chemists into thinking that nitrogenous nutrients must be assimilated without chemical change. In fact no one had been more deluded than Liebig himself! New analyses had shown that the nitrogen content of fibrin was higher than that of albumen and Liebig declared that there might well be a rearrangement of the elements of nutrient substances to form the matter of blood and tissues.

Liebig's attack on Mulder and the proteine theory destroyed the basis of his own theory of nutrition. He had asserted that non-nitrogenous foods did not take part in the formation of animal tissues and this assertion was based on his acceptance of the chemical identity of plant and animal albuminous matter. By 1847 Liebig became discouraged about the possibility of further developing his theory of nutrition and the third edition of the *Animal Chemistry* remained incomplete. Indeed Liebig began to point out the lack of detailed information in animal chemistry. There were so many intermediate steps in the processes of nutrition, metabolism and excretion and so few were thoroughly understood. In 1848 Liebig wrote to Wöhler, '... it is truly amazing how little is really known about the animal substances.' By this time Liebig and his students had turned away from elementary analyses in order to try to imitate some of the intermediate stages in animal functions and thus discover more about them. They studied the decomposition products of albumen and other substances and their work paved the way for later studies of the proteins, but it did not lead to any quick deductions about physiological processes. Knowledge of these was only slowly developed over a long period by those physiologists who followed Liebig's methods of investigation, particularly with respect to the quantitative estimation of rates of metabolic change. Liebig's suggestion that the rate of tissue metabolism could be measured by means of urinary nitrogen appealed to many of his followers because this seemed to offer a direct approach to the most fundamental vital processes.

About 1848 Friedrich Theodore Frerichs, professor of pathology at Göttingen, made some experiments to measure the basic rate of metamorphosis of tissues necessary to maintain the normal functions in animals. For this purpose he determined the rate of secretion of urea by the animals when fasting. When he discovered that this was only a small fraction of the rate for an animal on a nitrogenous diet he concluded in opposition to Liebig that the excess of nitrogenous nutrients over those required for the basic rate is directly oxidised in the blood in the same way that non-nitrogenous substances are. During the next twenty years this view was to compete with that of Liebig and perhaps the most influential piece of work done in order to support Frerichs' theory against Liebig's was that of F. Bidder and C. Schmidt, published in 1852. These workers investigated the total exchange of compounds and elements by all paths between an animal and its environment. They accepted the possibility of the direct oxidation of nitrogenous foods as suggested by Frerichs and called the process *Luxusconsumtion*.

In his attempts to supply a chemistry of animal functions Liebig's real weakness was his ignorance of physiology and anatomy. Theodor Bischoff, whose enthusiastic reception of the *Animal Chemistry* we have already noted, thought that he could supply this lack of knowledge. He therefore went to Giessen and joined forces with Liebig in attempts to provide the necessary experimental evidence to refute Frerichs' suggestion. It was first necessary to develop appropriate techniques and Bischoff spent several years studying the effects of variations in the quantity and type of food on the rate of urea excretion by dogs. In 1851 Liebig described a new and more accurate method of estimating urea in urine by titration with mercuric nitrate and Bischoff used this method in experiments on dogs designed to perfect the techniques both of measuring and controlling the food intake and the excretions. The object of Bischoff's work was to substantiate Liebig's theory that all nitrogenous food must first be converted into muscle tissue and all excreted nitrogen comes from the metamorphosis of these tissues. Should Frerichs' view of the direct oxidation of nitrogenous matters in the blood be found to be correct, Liebig's theory of nutrition would inevitably be shown to be wrong and all hopes of measuring the rate of metabolism in terms of urinary nitrogen would be lost. By 1853 Bischoff thought he could show that the views of Frerichs, Bidder and Schmidt were unlikely because when the amount of protein in the diet was increased progressively, the increases did not appear entirely in the urine but a proportion was always retained in the body.

Attempts to re-assess Liebig's theories

Following Liebig's procedure, Bischoff used the elementary compositions to calculate the amount of nitrogen in urea and in the tissues of meat. He found that more than one-third of the nitrogen in the food remained unaccounted for and concluded that there must be other ways in which this element is excreted. His results seemed to invalidate Liebig's theory, for even if his idea of muscle metabolism were correct there seemed to be no simple way of measuring the exchange of nitrogen. In his experiments Bischoff had weighed his dogs and had assumed that when no change in weight could be detected there was no change in the composition of their tissues either. This might not be true, for a gain in muscle tissue in an emaciated animal might be compensated by a loss of water. In 1857, Carl Voit a young assistant of Bischoff carried out a set of experiments in which the nitrogen in the food and in the excretions balanced and in some other tests he found *more* nitrogen in the urine than in the food. These results helped to restore confidence in the idea that urinary nitrogen might be a measure of muscle metabolism. Under Bischoff's direction therefore, Voit began a further set of experiments to test the earlier conclusions.

Once again using dogs, Voit measured the urea present in the urine when the animals were fasting and when they were given carefully graded amounts of nitrogenous food. He also used mixed diets of nitrogenous food and fats or carbohydrates in which he kept one type constant whilst varying the other. Then by means of calculations based on elementary analyses and on the weights of food, the excretions and the animal itself, Bischoff and Voit estimated the amount of protein assimilated and decomposed together with the weights of water and fats gained or lost in each case. The results of this extremely detailed study were published in 1860 in a book entitled '*The Laws of Nutrition of Carnivorous Animals, Established by New Investigations*' which represents the pinnacle of all attempts to base a science of nutrition on Liebig's principles.

In this book Bischoff and Voit accepted the idea that the decomposition of muscular tissue was the sole source of animal energy. They followed Liebig's suggestion that nitrogenous foods form first the organic constituents of the blood, then tissues, and are from that stage decomposed to yield urea, muscular energy and animal heat. Liebig's theories were in fact treated with great brilliance and the authors believed that they had firmly established them as inevitable conclusions from physiological experiments. But it was the conception of physiology itself which was at fault, for conclusions were drawn about the internal changes occurring within the animal from observations of the gross exchanges with the surroundings and in the end continued experimentation of the kind begun by Liebig was to lead to the final repudiation of his theories.

Voit himself took the first steps along the road to the abandonment of Liebig's theory of nutrition when he found that there is no significant difference between the rate of excretion of nitrogen in a resting dog and one which is made to run intermittently on a treadmill. Rather than discard Liebig's view that urinary nitrogen is proportional to work done, Voit tried to patch up the difficulty by suggesting that in periods of rest the metamorphosis of tissue continued and the resulting muscular energy instead of being expended was stored as an electric potential of the surface of the muscle. When the muscle later contracted it was this electric potential which provided the mechanical energy. As experimental work increased it became evident that the situation was more complex than Liebig had suggested. It was not possible in practice to treat the reactions leading to the metamorphosis of tissue in isolation from those producing animal heat, nor was it true to say that muscular tissue was derived entirely from the nitrogenous components of the food. Both types of food took part in all aspects of nutrition.

Moritz Traube, the German chemist who had been a pupil of Liebig but had reluctantly given up his academic career to take over the family wine business at Ratibor, was the first to make a definite break with Liebig's theory of nutrition.

Traube thought there were several reasons for believing that the oxidation of non-nitrogenous matter yielded muscular energy as well as animal heat. Draught animals for example, eat little protein yet the work they perform would, on Liebig's theory, require large quantities of nitrogenous food. Again, strenuous work results in increased respiration and the consequent 'burning' of non-nitrogenous food material, whereas the exposure to cold does not, yet in this case the animal must produce increased amounts of heat. Traube thought that such observations led to the conclusion that the oxidation of non-nitrogenous matter yielded mechanical energy and he suggested that Voit's results served to confirm this. It would seem unreasonable to try to accommodate these observations to fit Liebig's physiology and in any case, since heat and energy were recognised to be equivalent it seemed more reasonable to suppose that the food which supplied one might also supply the other.

About 1865 A. Fick, a Swiss physiologist and J. Wislicenus, a German chemist, made an experiment on themselves to test Liebig's assumptions that muscular energy was derived from the metamorphosis of tissues and that the extent to which this had occurred could be measured directly in terms of urinary nitrogen. They climbed a mountain in the Alps, carefully calculating the mechanical work done in the process and comparing it with the amount of nitrogen appearing in their urine during the same period. They found however that the effort needed to execute the climb was very much greater than could be reasonably thought to have come from the small amount of protein oxidised as indicated by the urinary nitrogen. To further substantiate their results, Fick and Wislicenus also pointed out that many fast moving or powerful animals (they cited goats, chamois and gazelles), took in little protein but plenty of carbohydrates. Thus it seemed that muscular power depended upon the oxidation of carbohydrates as well as nitrogenous foods.

Edward Frankland, the English chemist, considered that although this conclusion was acceptable the evidence for it supplied by Fick and Wislicenus was not precise enough to make it certain. In an attempt to supply the necessary precision therefore, Frankland determined the heats of combustion of urea, hippuric acid, albumen, fats and muscle tissues. The results of these, the first determinations of heats of combustion of foodstuffs, supported the earlier work of Fick and Wislicenus for they showed that scarcely one-fifth of the total energy needed for the work done could have come from the oxidation of muscle tissues. Frankland therefore concluded, in opposition to Liebig, that muscular power was derived *chiefly* from non-nitrogenous material in the blood. This calorimetric work begun by Frankland was developed until by 1893 it could be shown quantitatively that all the heat released in the resting animal was derived from the oxidation of its food.

Liebig on the defensive

Liebig himself took little part in the experimental work designed to investigate his theory of nutrition and animal heat, but growing opposition to his views led him in 1870 to write a long treatise answering the criticisms. He was now forced to admit that the measurement of urea in the urine was not a satisfactory method of estimating muscular work and he allowed that some protein might be oxidised directly in the blood during respiration. Despite this he stuck to the notion that the metamorphosis of muscular tissue is the only source of work and gave as evidence for this that an animal cannot live on an entirely non-nitrogenous diet. Without acknowledging its source, Liebig used Voit's idea of the build-up of reserves of energy during periods of rest and suggested that this energy was stored in the compound creatine which he had shown to be present in all muscular tissue. In any case this idea had some commercial value as well for Liebig was able to use it in order to explain the food value of his famous meat extract.

In this treatise Liebig specifically criticised Voit's experiments and so initiated another vigorous dispute in which Liebig now nearing the end of his remarkable career, found himself on the defensive against the attacks of Voit, as Berzelius had earlier been against those of Liebig. Voit asserted that Liebig's theory of nutrition had either been definitely shown to be false or at least unlikely. He declared that the experiments clearly indicated that there was *no* metamorphosis of tissue in the sense suggested by Liebig—decomposition through muscular work separately from the respiratory process. The creatine theory recently proposed was pure speculation and although it was true that many of the ideas expressed in the *Animal Chemistry* had also been speculative, these had led to investigations which were now making further progress on the same basis impossible. Unfortunately, said Voit, Liebig was still using the same ideas and methods which he had established twenty-five years earlier and he should now revise or relinquish them. The dispute between Liebig and Voit became very bitter and it reminds us of the earlier quarrel between Berzelius and Liebig. In each case the younger man was presenting work which he regarded as a logical extension of that of the older. In each case the younger man hoped that his revisions and innovations would be accepted by the older chemist in the cause of scientific progress, and in each case the younger man was instead confronted by strong opposition and all the weight of authority commanded by the older man. If we need to be reminded that science is a human activity and that those engaged in the pursuit of truth are endowed with all the human qualities and failings, we shall find all that is necessary in the story of Liebig's *Animal Chemistry* and the controversies it aroused during the second half of the nineteenth century.

Chapter 8
Contributions from Physical Chemistry

ALTHOUGH BIOCHEMISTRY is concerned with life-functions and is therefore made up in the main of complex organic reactions and substances involved in the chemistry of the living cell, it has nevertheless been necessary to borrow theories and methods from physical chemistry in order to understand the mechanisms of life-processes. The rigid division of natural knowledge into separate disciplines may have proved convenient but it is always artificial; there are few sharp boundaries in science, and biochemistry offers an excellent example of an interdisciplinary study. Since biological systems function by means of complex reactions controlled by physical mechanisms, it is necessary to consider some of the relevant parts of physical chemistry in order to understand the rise of biochemistry itself.

Helmholtz and the conservation of energy

Hermann von Helmholtz, the great German physicist and physiologist, was attracted by the major problems facing the physiologists in the nineteenth century—spontaneous generation for example, and the source of energy from which animal heat is derived. He applied the concept of the conservation of energy to animal life in a way which was to have far-reaching consequences on the direction which physiological research was to take in the later part of the century. It was in 1847 that he published his famous paper on this subject, after a long and careful study which had been initiated in part by assertions made by Liebig in the *Animal Chemistry*.

German thought in the early nineteenth century was strongly influenced by the metaphysical philosophy of Hegel who held that Nature could not be wholly rational. This outlook left the way open for a concept of vital force which had found expression amongst German natural philosophers in the form of the doctrines of *Naturphilosophie*. Helmholtz could not agree with such a view and in particular he could not accept the animistic approach which attributed to every living organism the internal power of perpetual motion as an essential attribute of its vitality. In his paper of 1847, Helmholtz gave strong reasons for rejecting all such ideas, for whatever the cause of vital force might be, he was certain that it did not include the performance of work without using up energy as some of the vitalists seem to have thought.

Even in the late eighteenth century Davy and Rumford had indicated the inter-relation between heat and work, whilst more recently J. P. Joule and J. R. Mayer had re-stated it in quantitative terms with experimental evidence to support it. These men, along with Lord Kelvin and Helmholtz himself, must be regarded as the founders of the law of conservation of energy, but it is Helmholtz who is of special importance to animal chemistry because he saw that the same law must apply to the living animal organism.

In 1845, three years after the appearance of Liebig's *Animal Chemistry*, Helmholtz pointed out that Liebig had posed one of the most fundamental physiological questions in the clearest terms. This was the question whether organic life resulted from the effects of a 'self-propagating, purposeful force' or whether the same forces which operate in inanimate nature also act in living bodies although in a modified form. Liebig had tried to show how physiological phenomena could be derived from physical and chemical laws and had raised the question 'whether or not the mechanical force and heat created in the organism can be derived entirely from chemical transformations'. It will be remembered that Liebig had suggested that an animal derives its muscular energy from the 'metamorphosis of tissues' and its heat from the oxidation of those parts of its food which consist chiefly of carbon and hydrogen. In both cases these chemical changes were said to be brought about by the action of oxygen taken in during respiration.

If Liebig's physiology were correct, Helmholtz argued that it ought to be possible to produce detectable changes in the chemical composition of muscular tissue by causing repeated, strong contractions in it. He therefore set up an experiment using a frog, in which by intermittent electric currents he could stimulate one muscle of a pair whilst the other remained at rest. After carrying this to the point of exhaustion he proceeded to extract soluble matter from each of these two muscles and as a result was able to show that the alcohol-soluble matter increased in the exhausted muscle whilst at the same time the water-soluble matter decreased. Although he was not able to identify the chemical substances which were the cause of this change, Helmholtz took this result as confirmation of chemical transformations in the muscle and for a long time his experiment was regarded as the best direct evidence that such transformations really occurred.

The problem of animal heat and the contention that it arose entirely from the oxidation of nutrient matter within the animal body was one of the central issues raised in Liebig's *Animal Chemistry*. As we have seen it was a problem of long standing. About 1780 Lavoisier and Laplace had investigated the relationship between animal heat and combustion and in 1824 two French physicists, Pierre Dulong and César Despretz had independently attempted to measure again the heat evolved by an animal and to correlate this with its respiratory exchanges. In their experiments both these workers had placed an animal in a vessel surrounded by water and had circulated air from an aspirator through the vessel into a second vessel in which the chemical analysis of the respired air could be carried out. From the volumes of oxygen absorbed and carbon dioxide evolved the predicted evolution of heat could be calculated, but both men found that the water took up only 50 per cent to 70 per cent of this calculated quantity. Even when the heat suspected to be evolved by the combustion of hydrogen to form water was added, they still could not account for more than 90 per cent of the total predicted value. As a result of these discrepancies Despretz suggested that although respiration was the principal source of animal heat, the motion of the blood, the reactions of assimilation and friction in the body must account for the rest. Dulong, on the other hand, concluded that respiration alone was insufficient to account for the heat losses which animals undergo and that there must be another source of animal heat which might never be known.

Johannes Müller weighed these opposing views and came to the conclusion that the deficit of animal heat must be made up from some *nervous* source. He pointed out that there was no proof of the formation of water during respiration—this was merely an hypothesis introduced by Lavoisier and Laplace. Liebig however felt sure that the whole of the heat produced by an animal must be accountable in chemical terms and he looked for new ways to achieve this. In 1845 he used a higher value for the heat of combustion of hydrogen derived from more recent experiments and he also suggested that the direct measurement of the heat of combustion of carbon was unreliable. Instead Liebig deduced the heat of combustion for carbon from those for simple compounds such as olefiant gas, alcohol and ether by subtracting the part due to combustion of hydrogen in each case. With these new values he went on to recalculate Dulong and Despretz's results and now found ratios for calculated heat of combustion to measured animal heat ranging from 0.83 to 1.04. This then allowed him to conclude in accordance with his convictions that the combustion of carbon and hydrogen in the body did in fact account for all the heat actually liberated in the body. Although superficially this appeared to strengthen the chemical theory of animal heat, it was really due to Liebig's firm belief that his ideas must be right that he chose to take these steps to make the experimental observations fit in with his scheme. We may question the validity of such an exercise from the scientific point of view and in fact Liebig himself was well aware that the assumptions he had made in order to bring the figures into line with his theory were open to question.

The key to the problem was provided by Helmholtz who showed that *neither* Dulong and Despretz's first results *nor* the new interpretation of them given by Liebig was adequate to show that the whole animal heat must be derived from oxidation. All these workers had assumed that the heats of combustion of the complex organic substances found in nutrients were the same as the sum of the total heat evolved during the combustion of the separate elements carbon and hydrogen of which they are composed. After making some experiments on the heats of combustion of substances similar to the major nutrients Helmholtz concluded that this assumption was simply not true. The quantity of heat evolved during the combustion of an organic compound can be much greater or much less than the sum of all the heat evolved by the combustion of equivalent weights of its constituent elements. Thus heats of combustion cannot be calculated. It followed from this that the results of Dulong and Despretz did not constitute an obstacle to the chemical theory of animal heat. Helmholtz was not able to prove this theory any more than Liebig had been, but like Liebig, he felt certain that the intrinsic energy of the nutrient matter must be the source of animal heat since there was no real evidence of any other source and he believed that his latest ideas removed some of the evidence which had appeared to be against this.

Thus, Helmholtz recognised that the evolution of animal heat depended upon the conservation of energy and that it was a special case of a more general law. By 1847 he had come to believe that Dulong and Despretz's results, far from disproving Liebig's theory of animal heat, actually supported it. His famous paper, in which physiological problems were treated according to the latest ideas about the conservation of energy, marked him out as a pioneer, since physicists themselves had failed to grasp the significance of their work for physiology. After this time it was tacitly assumed that the conservation of energy within the animal body was axiomatic and that the source of heat and work in any animal could be traced entirely to the oxidation of nutrients. Nevertheless, the release of this heat and energy was known to be under the control of the nerves and the questions still remained to be answered concerning the relationship between the intensity of the nervous impulse and the degree of force exerted by the contracting muscle.

Electricity and Life

In 1791 Luigi Galvani, who was a physiologist working in the University of Bologna, noticed the contractions in a frog's leg which he had prepared for dissection. Noting that the contractions occurred when a scalpel held by its metal parts came into contact with the crural nerves of the frog, Galvani concluded after many experiments that an electric current passed in the frog's leg when its nerve and muscle formed a complete circuit which was earthed by the operator. From his results he came to the conclusion that there was a store of 'animal electricity' in the muscles of the frog's leg which was discharged when two dissimilar metals were brought into contact with the nerves, the muscle and each other. In this way Galvani had discovered both the possibility of an electric *current* and the fact that this was connected with the function of the nerves. The results of his experiments were described as he had obtained them, with little concern for orderly arrangement, but he made several useful suggestions including the idea that electric currents might be found valuable in the treatment of certain diseases such as rheumatism, epilepsy and paralysis. Galvani thought that animal electricity was the essence of 'animal spirits', but he recognised that it was identical with all other forms of electricity and capable of producing all the well-known electrical effects. Galvani's nephew, Giovanni Aldini, professor of physics at Bologna and something of a showman, worked hard to secure popular recognition for his uncle's discoveries and went to great lengths to show how electricity might be employed in medicine.

Animal electricity was at once taken up by Alessandro Volta, professor of physics at Pavia. Volta set out to disentangle Galvani's observations, beginning with the concept of a reservoir of animal electricity which could be released by the application of dissimilar metals to moist animal surfaces such as the tongue and even the eyes. In each case he found that the appropriate sense organs were stimulated. Thus, an acid taste was produced on the tongue, whilst flashes of light were seen when the metal plates were touched on the eye. Since the same plates also caused muscular contractions, it followed that the flow of electricity through the nerves gave rise in each case to the proper function of the nerves, whether they were sensory or motor. This led the physiologist Müller to the law of specific nerve energies, first stated in 1840. The effects of stimulating a nerve are independent of the form of stimulation but depend only upon the organ to which the nerve is connected. Berzelius, as we have seen, also accepted the idea that the driving force of life-functions was probably located in the nervous system.

In the early nineteenth century little was known about the structure and functions of the nerves and these electrical discoveries led to a general extension of interest in the nervous system. In this country Sir Charles Bell, famous as an anatomist and surgeon, traced the courses of the more important nerves, distinguishing between the two main classes of motor and sensory nerves. Bell did not like animal experimentation and it therefore fell to François Magendie, the French anatomist and physiologist, to distinguish the posterior (sensory) roots of the nerves from their anterior (motor) roots. At St George's Hospital in London, Everard Home and Sir Benjamin Brodie examined the structures of the nerves supplying the lungs, limbs and viscera. Electric currents were known to cause coagulation of the blood albumen and it was suggested that animal electricity might be the cause of the secretions. Brodie found that when a dog was poisoned with arsenic there was an excessive secretion of mucus in the stomach, but if the nerves of the stomach were divided before the arsenic was given the mucus secretion did not occur. In this case at least it appeared, the secretion of mucus was controlled by the nerves. Home also investigated the control of animal heat by the nerves, as we have already seen.

The action of the nerves in controlling respiration and digestion was investigated by A. P. Wilson Philip who showed that when the nerves to the stomach and lungs were severed and the ends separated in a rabbit, digestion stopped and the stomach remained full of partially digested food. At the same time the breathing became laboured. In both cases he found that by applying the terminals of the electric battery to the organs, normal healthy functions could be restored. Philip concluded that the effect of dividing the nerves to a vital organ was to derange the powers on which its healthy structure and functions depended, whilst the effect of the electric current was to restore these powers.

It had been suggested that mechanical irritation of the nerves was just as effective in stimulating them into action as was the electrical effect. In order to test this assumption Philip took three rabbits, fed them all at the same time and then severed the nerves to the stomach in two of them. In one of these he tied the severed ends of the nerve to a muscle so that the nerve would be in continual mechanical motion. After ten hours one of the rabbits died and the others were then killed, their stomachs were removed and opened. Brodie was asked to examine their contents, but though he could pick out the healthy stomach he could not distinguish the other two. From this Philip concluded that mechanical irritation had little or no effect on the action of the nerves and was certainly much less effective than electricity or galvanism.

One consequence of such observations was that current electricity began to be used as a means of treatment for certain kinds of nervous disorders. Since the time of Galvani and Aldini static electricity had been used in medical treatment, but Philip now applied both electric shocks and currents to patients suffering from chronic respiratory diseases. He would place one terminal from a battery on the nape of the neck and the other on the pit of the stomach. Such methods, introduced about 1815, rapidly grew in popularity. In France, the physician Duchenne introduced electrical methods both for diagnosis and treatment; he introduced a procedure called 'electropuncture' of the muscles. At Guy's Hospital in London, a special 'electrifying room' was set up about 1836 under the direction of Golding Bird, a well-known London physician. The equipment of this room included voltaic batteries, Leyden jars and an induction coil with an automatic make and break device designed by Bird himself. With this apparatus Bird could administer electrical treatment ranging from strong shocks to feeble continuous currents. In one method the patient was placed on an insulated stool and was connected to the positive pole of the battery by attaching a metal terminal to his skin. In this state he was said to be in an 'electric bath' and sparks could be drawn from his spine! This treatment proved to be very effective for patients suffering from St Vitus's Dance and some forms of paralysis. The electric currents seemed to act solely by stimulating dormant nervous functions.

Whilst animal electricity was finding applications in medicine, some very interesting quantitative experiments on the relationship between electric impulses and muscular contractions were being carried out in Pisa by Carlo Matteucci, an Italian physiologist. Beginning about 1836 with an investigation of the electric organ of the torpedo, a Mediterranean fish capable of producing a substantial electric shock, Matteucci went on to extend Galvani's work on animal electricity in the frog. He built up 'piles' from frogs' leg muscles, arranged so that the inner surfaces of one element were in contact with the outer parts of the next. This arrangement can be regarded as analogous to the voltaic pile and it was found to produce electric currents. As a sensitive detector of these currents, which were very feeble, Matteucci used a specially prepared frog in which the thigh muscles were left attached to the lower half of the spinal cord by the crural nerves only. When the claws of this 'galvanoscopic frog' were placed in contact with opposite ends of the pile of muscle elements, it contracted vigorously. By taking his muscle.elements from frogs which had been kept at different temperatures and in different states of nutrition, Matteucci found that the intensity of the contractions of the galvanoscopic frog corresponded with the state of vitality in the muscle elements used. He suggested that it might be possible to correlate the forces of muscular contraction with the intensity of electric discharges through the nerves supplying the muscles and devised an apparatus in which this could be attempted. It turned out that there is in fact such a direct relationship and Matteucci claimed that his experiments '... sufficiently demonstrated, that the electrophysiological effect is proportional to the intensity of the current.'

From results such as these electrophysiologists felt that they were coming very close to quantifying the vital force itself. Vital force was to physiology as affinity was to chemistry and attraction to physics. Experiments on living animals indicated that there is a direct relationship between electricity and nervous energy; it had been shown that the vital organs such as the lungs, heart and stomach are supplied by ganglionic nerves stemming from all parts of the brain and spinal marrow. Such organs were therefore influenced by changes in almost any part of the body. Prout had suggested that the whole digestive system might be considered to function on galvanic principles and Matteucci's work indicated that, at least in the case of muscular contractions, there is a *quantitative* relationship between the force exerted by the muscle and the intensity of the electrical stimulus. In fact his experiments indicated that the quantity of work which could be obtained by the discharge of a given quantity of electricity through the muscle was far greater than the work obtained when the same quantity of electricity was used in a physical machine. The living body could be regarded as an extremely efficient mechanical system.

Electrolysis and the behaviour of solutions

Volta's experiments on animal electricity soon led him to discard this concept and replace it by another—the idea of 'metallic electricity'. He showed that an electric current could be obtained from an arrangement of two dissimilar metals separated by a conducting solution (e.g. salt or dilute acid). Early in 1800 Volta wrote a letter to Sir Joseph Banks, President of the Royal Society, in which he described his invention of the electric 'pile'. Each unit of this device consisted of a disc of copper and a disc of tin or zinc separated by cardboard soaked in brine or dilute acid. A pile of such units was then built up with dissimilar metals of adjacent 'cells' of the pile in contact. This device was found to be capable of delivering a substantial electric shock, but more important it yielded a continuous supply of electricity—an electric current. Almost at once William Nicholson and Sir Anthony Carlisle together stumbled upon the phenomenon of electrolysis and within a year Humphry Davy, working in the laboratories of the Royal Institution in London, began experiments on electrolysis which culminated in his discoveries of sodium, potassium and the alkaline earth metals. These results showed that electrolysis could be used as an analytical tool and also made it clear that matter and electricity are related at the atomic level. Davy himself thought that the elementary particles only became charged when they came into *contact* with each other, whilst Berzelius held that the atoms were individually either positive or negative at all times. Both regarded these charges as the source of chemical affinity and Berzelius also suggested that when the charges were neutralised heat and light were released. He went on to develop a quantitative system of chemical explanation, based on the existence and interaction of these charges. This 'dualistic' system of chemistry was very successful when applied to inorganic compounds, but it was much more difficult to see how it could be used to explain complex organic reactions, although attempts were made to do so by means of the concept of organic radicals, first used by Lavoisier.

The term 'ion' to describe an electrically charged atom or radical was introduced by Faraday in 1834, when he showed that there is a quantitative relationship between the total current passing through the solution and the chemical equivalents of the ions carrying that current. Faraday stated the erroneous idea that each formula weight of an electrolyte must contain one equivalent of each ion, but he rightly thought that the *movement* of the ions in opposite directions gave rise to the flow of electricity through the solution.

Faraday's work laid the foundations for the study of electrolysis from the physicists' point of view, but speculation about the distribution of the ions in solution continued amongst physical chemists throughout the greater part of the nineteenth century. About 1857 Clausius suggested that a *small* proportion of the solute molecules in a solution of an electrolyte were dissociated into ions.

$$AB \rightleftharpoons A^+ + B^-$$

During electrolysis these ions were deposited at the electrodes and more solute molecules then dissociated so as to keep the proportion of ions present in the solution constant. This explanation accounted well enough for the observations of electrolysis and was accepted generally for the next thirty years until it was eventually displaced by Arrhenius' theory of almost complete ionisation, a concept made necessary by observations of the abnormal behaviour of electrolytes in solution with respect to vapour pressure, elevation of the boiling point, depression of the freezing point and osmotic pressure.

Osmosis

The transference of matter between the living cell and its environment is a process fundamental to all biological systems. Nutrition, respiration and excretion all involve such changes which require energy and are brought about by osmosis and diffusion through the cell membrane.

Even as early as 1748, the Abbé Nollet in Paris had shown that when a cylinder containing alcohol was closed with a membrane of bladder and then immersed in water, the membrane began to bulge outwards due to the passage of water into the alcohol. A process based on this observation was later used as a method of strengthening brandy. It was explained in terms of affinity by G. F. Parrot, professor of physics at Dorpat, but the first quantitative experiments on osmosis were made by H. Dutrochet, a French army doctor, about 1824.

Dutrochet confined his liquids in animal or vegetable membranes and investigated what he called endosmose as material passed through the membrane. He was most interested in the applications of this process in plants, but he pointed out that it could occur using mineral substances just as readily as it did in the physiological situation. In fact, he said, the more one came to know about physiology the more reason one had to believe that the phenomena of life are not essentially different from physical phenomena. Dutrochet explained osmosis in terms of an electric current set up between two liquids of different densities, placed in contact with each other. This current carried the less dense liquid towards the denser. In osmosis it is, of course, the difference in molecular concentration which causes the effect and Dutrochet was confusing this with electro-endosmose, a phenomenon which he also investigated.

The apparatus used by Dutrochet consisted of a bell-shaped glass vessel the open end of which was covered with a bladder. This was then filled with the solution and placed in a larger vessel containing water. A tube attached vertically to the osmometer measured the pressure set up. This was sometimes quite large; some of his sugar solutions produced a pressure equivalent to a column of mercury one metre high. Dutrochet observed the phenomenon of osmosis in both plant and animal cells and this led him to assert, '... The physiological connexions which I have established between plants and animals make it clear that there is but a single physiology, a general science dealing with the functions of living beings ...'

The first reliable measurements of osmotic pressure were made by the German botanist, Friedrich Pfeffer in 1867. The greatest difficulty to be overcome in this work was the production of a semi-permeable membrane which would withstand the high pressures involved and capable of allowing quantitative measurements to be made. This problem was solved by Pfeffer, as is well-known, by depositing a coherent layer of ferric ferrocyanide (Prussian blue) in the pores of an earthenware vessel. With this apparatus Pfeffer was able to show that the osmotic pressure of a solution is directly proportional to its concentration and to the Absolute temperature.

$$P \, \alpha \text{ conc. and } P \, \alpha \, T^\circ \, A.$$

About the same time the Dutch physical chemist J. H. Van't Hoff, confirmed these results and went on to extend his observations to solutions containing electrolytes. He found that a non-electrolyte in solution obeyed a relationship similar to the perfect gas equation, $pv = RT$, but an electrolyte always showed anomalous behaviour and it was necessary to introduce a factor 'i' into the right-hand side of this expression. Van't Hoff's 'i' was found to be approximately equal to the number of ions per formula weight for any strong electrolyte—a fact which lent support to Arrhenius' suggestion of almost complete ionisation in strong electrolytes.

When two solutions have the same osmotic pressure they are said to be istonic. This condition, which is of some importance in physiological chemistry, was demonstrated experimentally by Hugo de Vries in 1884 using plant cells which he observed under the microscope. In a healthy plant cell the cytoplasm is pressed tightly against the cell wall and the cell is said to be turgid, but if it is placed in a solution of higher osmotic pressure than that of its contents, water will pass out through the cell wall and the cytoplasm will shrink. This is called 'plasmolysis'. It is possible to prepare solutions which just fail to bring about this effect and series of such solutions containing different solutes can be prepared. These solutions all have the same osmotic pressure as the contents of the cell and they are said to be istonic with each other and with the cell contents. A similar effect was observed with red blood-corpuscles by Hamburger in 1886. It has been invaluable for the preparation of 'normal saline' solutions for use in blood transfusions. In such solutions the red blood-corpuscles neither become distended nor do they shrink because the osmotic forces on each side of their membranes are balanced.

Most of the fluids found in the animal body are either acidic or alkaline and this property is always an important feature in their chemical behaviour. Water itself is very slightly ionised, containing 10^{-7} gm. ions each of hydrogen (H^+) and hydroxyl

(OH⁻) ions per litre. Since these are the two ions responsible respectively for acidity and alkalinity and they are present in equal proportions in water it follows that water must be regarded as *neutral*. Any solution in which the concentration of hydrogen or hydroxyl ions is found to be 10^{-7} gm. ions per litre is likewise neutral, but if a solution contains any other concentration of these ions it must be either acidic or alkaline. Thus, a solution containing, say 10^{-5} gm. ions of hydrogen per litre is acidic, whilst one containing 10^{-9} gm. ions of hydrogen per litre is alkaline. Now these numbers expressing hydrogen ion concentration are inconvenient to use and it has therefore become standard practice to adopt the device suggested in 1909 by S. P. L. Sørensen, an enzyme chemist at the Carlsberg breweries. Sørensen was interested in the physical chemistry of the proteins, especially those which act as enzymes. Acidity or alkalinity was recognised as important for the stability of the proteins and for the *rate* of enzyme action, so Sørensen made a very thorough investigation of the effects of hydrogen ion concentration, paying particular attention to its accurate determination. He stated the hydrogen ion concentration as the reciprocal of its logarithm, calling the function the hydrogen ion exponent (or pH) of the solution.

$$pH = -\log_{10} [H^+]$$

The examples given above would then have pH values of five and nine respectively, numbers which are much more convenient to deal with. It is interesting to note that the physical concept of pH grew up within the experimental study of enzyme reaction rates. The one slight disadvantage of the system is that acid solution with high hydrogen ion concentrations have pH values *below* 7 and alkaline solutions whose hydrogen ion concentrations are low have pH values *above* 7.

In one section of his paper on the study of enzymes Sørensen gave an account of the preparation of buffer solutions for the accurate estimation of the pH of solutions. A buffer solution contains a weak acid or alkali together with one of its salts, mixed in such a manner that the ionic balance is self-adjusting with respect to H and OH ions. Sørensen carefully analysed the buffering ranges of the borates, citrates, phosphates, compounds still commonly used for this purpose, and glycine. At the end of his paper he gave examples of the application of buffer solutions to the determination of the optimum pH values for the reactions of the enzymes invertase, catalase and pepsin. This was the beginning of a large chapter of biochemistry and some of the most important advances in the subject have concerned the study of ionic equilibria. The behaviour of the blood, kidneys and the whole function of respiration depends upon it.

Sørensen proceeded to investigate the behaviour of proteins in solutions which also contained ions. He found that certain proteins such as serum albumen behaved as though they were series of dissociable complexes from which a variety of precipitates could be obtained the compositions of which depended upon salt content and pH of the solution. Other proteins, including haemoglobin, egg albumen, keratin, trypsin and pepsin behaved as stable units or single compounds. Thus proteins could be classified according to this distinction in their behaviour.

Thomas Graham; osmosis, diffusion and colloids

The phenomena of diffusion and osmosis in liquids were linked by Thomas Graham in a study of the behaviour of solutions when superimposed upon each other in layers caused by density differences, or separated from each other or the pure solvent by thin organic membranes. He noticed marked differences in behaviour between organic substances such as albumen and gelatin, and crystalline compounds such as sugars or mineral

Albrecht von Haller.
(1708 - 1777)

A. F. de Fourcroy.
(1755 - 1809)

William Prout.
(1785 - 1850)

Plate 5

Baron Jons Jakob Berzelius.
(1779 - 1848)

Justus von Liebig.
(1803 - 1873)

Johannes Müller.
(1801 - 1858)

Plate 6

Claude Bernard.
(1813 - 1878)

Henry Bence Jones.
(1813 - 1873)

Hermann L. F. von Helmholtz.
(1821 - 1894)

Johann L. W. Thudichum.
(1828 - 1901)

Plate 7

Ernst Felix I. Hoppe-Seyler.
(1825 - 1895)

Emil Fischer.
(1852 - 1919)

Sir Frederick Gowland Hopkins.
(1861 - 1947)

Plate 8 (Courtesy of 'Nature')

salts. In 1861 he introduced the terms 'colloid' to describe the former and 'crystalloid' to describe the latter.

Although a few isolated examples of metals and other mineral substances in the colloidal state were already known before Graham's time, it was he who laid the foundations of colloid chemistry in five important papers published in *Philosophical Transactions* between 1850 and 1864. In the first of these papers Graham dealt with the diffusion of liquids into each other without the use of a membrane. His method was to fill a glass phial with an aqueous solution of known density and to place it in a larger vessel full of pure water. After a measured interval of time the phial was carefully closed with a glass plate, removed from the larger vessel and its contents were analysed. In this way the rate of diffusion could be found and Graham was able to show that it was roughly proportional to the concentration of the original solution. He also found that the rate of diffusion increased with rise in temperature and he discovered groups of similar compounds whose diffusion rates were about the same when the absolute concentrations of their solutions were equal. Later he repeated these experiments using mixtures of salts and double salts in solution and in these cases he found that diffusion could effect a separation, indicating that the process might be used as a method of analysis. Graham pointed out the similarity between liquid and gaseous diffusion; each solute behaved as though the others were not present. The rate of diffusion from one solution to another was not affected by the presence of the solutes but was always approximately the same as the rate from either solution into pure water. As with gases at low pressures this only held good for dilute solutions and departures from normal behaviour were observed when concentrated solutions were used. Albumen in the form of egg white was found to have the slowest rate of diffusion, but neither albumen nor gelatin in the solution reduced appreciably the rate of diffusion of the salts. This led to a good separation of mineral compounds from these organic substances and Graham thought that it could be used as 'a delicate method of proximate analysis peculiarly adapted for animal fluids.'

In 1854, Graham turned to the investigation of osmosis. His first apparatus consisted of a glass tube and cover fitted to the mouth of a porous clay cylinder 5″ deep and 1.7″ in diameter. The cover was made of guttapercha and the glass tube 0·6″ in diameter, was 6″ long. This apparatus filled with a saline solution was placed in a large outer vessel full of distilled water the level of which was adjusted to the height of the solution inside the tube so as to eliminate the effects of differences in hydrostatic pressure. In this apparatus the porous pot acted as a membrane and Graham found that when salt solution was separated from pure water *two* currents were set up. Salt diffused slowly out into the water whilst at the same time pure water passed through the porous pot into the salt solution under the influence of osmotic pressure. Similar results were obtained when the two liquids were separated by a membrane of ox bladder, although in this case it was necessary to employ very dilute solutions of the order of one per cent. It is interesting to note that Graham followed Dutrochet in trying to explain the phenomena electrolytically. He suggested that the hydrogen ion could travel through an aqueous medium by attaching itself to a water molecule to form a complex $H_{(m+1)}O_m$ (O=8). In osmometers using animal membranes, Graham took only the inner coat of ox bladder, supported on a thin plate of zinc which was painted or varnished to prevent the action of acids upon it. The membrane was tied tightly over the open end of a glass bell 3″ in diameter, the vessel so formed was then filled with the saline solution and immersed in a larger vessel of pure water. During the process of osmosis it was found that the membrane gradually lost its substance until its weight had decreased by 20 per cent to 40 per cent.

In 1861 Graham applied the method of liquid diffusion to the analysis of aqueous solutions containing mixtures of organic and crystalline mineral substances. The process by which these substances could be separated through parchment membranes was called *dialysis* and Graham distinguished between organic substances like albumen and gelatin which would not pass through the membrane—the colloids, and those substances, mostly mineral salts which would pass through—the crystalloids. The situation created could be seen to apply to animal cells and Graham suggested that the energy which he knew to be involved in the process might be the source of vital force.

'The colloidal ... a dynamic state of matter ... possesses ENERGIA. It may be looked upon as the probable primary source of the force appearing in the phenomena of vitality.' Graham indicated his belief that osmosis played an important part in all the vital functions. In life all the parts of animals undergo a continuous process of decomposition and renewal and so any losses in the membranes which might occur as a result of osmosis is made up by the life-functions of the animal. '... chemical osmose ... is peculiarly excited by dilute saline solutions such as the animal juices really are ... the alkaline or acid property which these fluids possess is another most favourable condition for their action on membranes.' But Graham also saw osmotic pressure as the means by which the living cell converted chemical energy into mechanical energy. Little or nothing was known about the mechanism by which an animal derived its muscular energy from its food or from the oxidation of its tissues. 'May it not be hoped, therefore,' said Graham, 'to find in the osmotic injection of fluids the deficient link which intervenes between chemical decomposition and muscular movement?' In any case in the microscopic cells osmotic pressure should attain its maximum since it is entirely dependent upon surface area. Here then we have yet another plausible explanation of the origin of vital force.

Optical activity—characteristic property of natural substances

One of the more striking physical properties of natural substances is the capacity of rotating the plane of plane polarised light. J. B. Biot, professor of physics at the Collège de France, first recognised this property in quartz *crystals* and in 1815 showed that liquid organic substances such as oils of turpentine, laurel and lemon were also optically active. Three years later he went on to show that solutions of certain sugars had optical rotatory properties. In such cases it seemed, the property must be connected with asymmetry in the *molecules* themselves, since they were arranged at random in the liquid and by the fact that they were in continual motion, the optical activity could in no way be due to any stable arrangement of the molecules as in the quartz crystal.

Biot's work inspired Louis Pasteur who began work on the subject using tartaric acid. He first prepared crystals of this compound and some of its derivatives which he proceeded to separate by careful inspection under a simple microscope. There were two types of tartaric acid crystal of which one was the mirror image of the other and when he found that in solution these rotated the plane of polarised light in opposite directions, Pasteur concluded that the *molecules* of tartaric acid could exist in two opposite configurations of which one is the mirror image of the other.

Ten years later Pasteur discovered another method of separating optical isomers by allowing certain types of simple organisms such as plant moulds to grow in a solution containing both optical isomers. He found that the solution became slowly optically active as the mould absorbed one optical isomer in preference to the other. Some asymmetry in the living organism caused it to act on only one isomer of the racemic pair, leaving the other more or less untouched. Pasteur suggested that this was the one clear line of demarcation which distinguished organised from inanimate matter—i.e. organised matter contains and gives rise to asymmetrical molecules of which one is preferred to the other, whereas non-living matter normally contains the two members of a pair of optical isomers in equal numbers. Hence by ordinary synthetical chemistry in the laboratory it is impossible to prepare separate optically active isomers, but in organised matter such isomers are commonly found in isolation.

The problem facing organic chemists was to correlate the optical activity of organic compounds with their molecular configurations. In 1862, A. M. Butlerov a Russian chemist who made important contributions to the study of molecular structures, suggested the possibility that the four valency bonds of the carbon atom might be directed towards the corners of a regular tetrahedron. This idea was revived five years later by Kekulé in an attempt to visualise the structures of acetylene and hydrogen cyanide, both of which have triple bonds in their molecules. Then about 1872 Van't Hoff, the Dutch physical chemist, worked with Kekulé and so became familiar with the idea of tetrahedral arrangement for the valency bonds of carbon. Using this principle he published a paper in 1874 in which he indicated in specific cases how the tetrahedral arrangement of four different groups about a central carbon atom would give rise to molecular asymmetry and hence optical activity.

Van't Hoff pointed out that organic constitutional formulae, as they were then written in two dimensions, were increasingly unable to account for certain cases of isomerism. He suggested that this arose from the need to extend the formulae into three dimensions and to imagine the relative positions of the groups and atoms with respect to each other *in space*. Starting with the methane molecule, the simplest organic molecule, he considered the result of replacing in turn each of the four hydrogen atoms by a different group, to give the general formula CRR'R"R'''. If these four groups were then placed at the corners of a regular tetrahedron, two isomers would be formed of which one is the mirror image of the other.

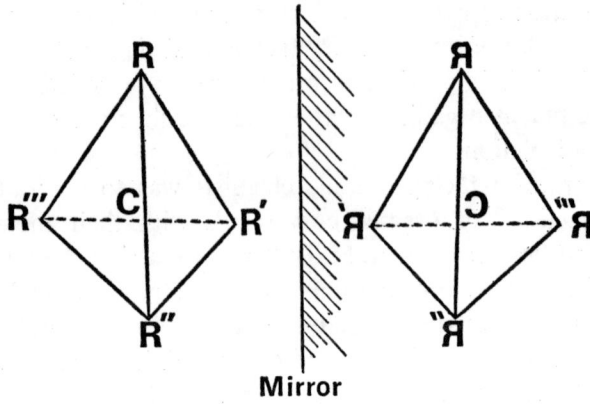

Mirror

Many examples of organic compounds which show optical activity were given by Van't Hoff. In each he was able to point out an asymmetric carbon atom and by searching amongst the known formulae of organic substances he was able to indicate in which compounds optical activity might be found as well as those in which it could not occur. He indicated the limits of optical activity by naming the simplest alcohols, acids, hydrocarbons, etc., which could show this property and pointing out that normal (i.e. straight chain) hydrocarbons, alcohols and acids, would be excluded from showing optical activity. Van't Hoff's paper was written from the standpoint of the practical organic chemist. He tried to account both for the presence of optical rotatory power and for its absence in cases in which it would be expected to occur. Thus when a mixture of equal parts of two optical isomers was formed, no optical activity would be observed and in cases such as tartaric acid in which each molecule contained two asymmetric carbon atoms one half of the molecule might exactly compensate for optical activity in the other half. Van't Hoff was asking chemists to use their imagination to picture the positions of the atoms and groups within the molecules and it was for this that he was strongly criticised by the German chemist Kolbe, who said that in Van't Hoff's paper, '... the play of the imagination completely forsakes the solid ground of fact and is quite incomprehensible to the sober chemist.'

Nevertheless, the discovery of molecular asymmetry was at the same time made by a French chemist, Joseph Achille Le Bel, who had been a pupil of Pasteur. Le Bel arrived at the concept from geometrical principles which led him to say, 'In general then ... if a body is derived from the original type MA_4 by the substitution three different atoms or radicals for A, its molecules will be asymmetric, and it will have rotatory power.' Le Bel considered the phenomenon of optical activity amongst carboxylic acids, such as lactic, tartaric, malic, valeric etc., and also in the sugar series.

The concept of molecular asymmetry as a means of explaining optical activity has been of immense value to all those organic chemists who have turned their attention to natural substances. It has enabled chemists to decide with a high degree of probability, between possible structural formulae for very complex substances. In particular, amongst sugars, alkaloids, proteins and other complex molecules the concept has proved indispensible as an aid in elucidating the structures.

Biophysics

Most of the ideas discussed in this chapter can be said to belong rather to physics than to chemistry and indeed it is now recognised that physics has much to offer the biochemist by way of experimental methods and tools for measurement. The introduction of the electron microscope and ultracentrifuge have greatly aided the study of medical problems related to virus diseases, cancer, heart diseases etc., and in addition many new situations now arise, brought about by space research and the use of nuclear power which reveal man in relation to unusual or novel aspects of his environment. This all leads to the relatively recent subject of *biophysics*.

There are four main branches of biophysics as it is known at present. Classical biophysics applies physical mechanisms to explain physiological phenomena such as the transmission of nerve impulses, the mechanism of muscle action and of the senses and why living cells can ingest some substances but excrete others. More recently changes in the normal physiological functions with changes in the physical environment (e.g. under high 'g') have been studied particularly in the United States in connection with space research. Mathematical or theoretical biophysics is the study of the behaviour of living organisms on the basis of mathematical physics. Thermodynamics and statistical mechanics are used to construct mathematical models which will simulate biological functions. In molecular biophysics large molecules such as viruses, nucleic acids and some proteins, are studied using electron microscope techniques, the ultracentrifuge and X-ray diffraction. Lastly in radiation biophysics the response of living organisms to ionising radiations (α, β, γ rays, X-rays and ultraviolet light) is studied so as to lead to a better understanding of the causes of mutations in cells and of the changes occurring at death. In these ways the physicist too plays an important part in developing explanations of vital phenomena; a fuller consideration of some physical techniques used in biochemistry is given below in Chapter 12.

Chapter 9
Claude Bernard; Physiologist and Animal Chemist

CLAUDE BERNARD was the most famous pupil of François Magendie, the French experimental physiologist whose principal criterion was to believe only what he could demonstrate for himself. Magendie was averse to all theories and would accept nothing which could not be experimentally verified and his influence was at first very strong upon Bernard who brought to animal chemistry the curiosity of the physiologist coupled with the methods of the exact sciences. Taking his cue from Magendie, Bernard conducted experiments on living animals; his training had quickened his powers of observation but he realised the need to break away from the purely empirical approach of his master. Bernard applied the results of his physiological experiments to the practice of medicine and the treatment of diseases. He introduced scientific methods to the study of the chemical phenomena of life, realising the need for rational hypotheses in order to construct a true science of physiological chemistry. In his most famous book, published in 1865, he outlined his ideas on this subject although he called it an *Introduction to Experimental Medicine*. In this book Bernard gave a general discussion of the moral issues attendant on animal experimentation and its justification. Whereas Magendie had been quite insensitive to the suffering caused to the animal during his experiments, Bernard was against causing animals unnecessary pain and was only prepared to accept vivisectional experiments which could be justified by their potential results. Later, in the last ten years of his life, he took up plant physiology with the intention of writing a general textbook of physiology. The first volume of this work appeared in 1872 but the project was not completed.

Bernard's scientific work began about 1843 with the study of a nerve in the face which activates a salivary gland—the *chorda tympani*. This nerve was found to take its origin from the facial nerve and was involved in facial paralysis. Bernard got his anatomy right but his physiological ideas on this subject were mostly wrong. He examined the saliva after this nerve had been cut and, failing to detect any difference he concluded that it had no influence on the secretion of saliva. Magendie had said that pinching the chorda tympani did not produce the sensation of pain and this indicated that it was not a sensory nerve and so was not concerned with taste. It was commonly accepted that the lingual nerve is the sensory nerve of the tongue, yet Bernard found that when the chorda tympani had been cut there was a reduction in sensitivity to the taste of dry citric acid placed on that side of the tongue. This led him to suggest that the chorda tympani, whilst not mainly responsible for the sensation of taste, must be accessory to it. He should have investigated the nerves of the tongue with greater care and should not have assumed that a dry substance placed on the tongue would not spread beyond the area to which the nerves had been severed. Further investigations led him to conclude that this nerve was purely motor and he rejected any suggestion that it carried sensory taste fibres.

Bernard's second false start was made in connection with digestion. Although he was well aware of the work done by Beaumont, he nevertheless concluded that the acid present in digestive juice was not hydrochloric but lactic because this was the only acid found in the stomachs of fasting dogs. Bernard distilled mixtures of lactic acid and common salt in aqueous solution, noting that the same stages were observed as in the distillation of gastric juice itself. He was right when he pointed out that if an acid reaction were *all* that was needed for digestion to proceed then *any* acid would do and he showed that the real digestive agent in gastric juice was the organic matter which was also present. This was found to be destroyed by heating to 85° or 90°C—enzymes are, of course, destroyed by heat. Bernard therefore subscribed to the mistake which had been made since the earliest experiments on digestion— that of assuming that the gastric juice is always present in the stomach. It was necessary, as others had found, to extract gastric juice from the stomach during digestion in order to determine its real composition.

In 1845 Bernard was awarded the prize for experimental physiology by the French Academy of Sciences. This was given for a paper in which he described the removal of the spinal accessory nerve by pulling it out through the foramen of the skull. This left the vocal chords apart and relaxed so that breathing was unimpaired but the animal became voiceless. In another case, when he severed the laryngeal nerves containing fibres from the vagus (the pneumogastric nerve), the vocal chords almost closed the glottis and the animal was in danger of suffocation. Bernard therefore concluded that these two nerves acted oppositely on the glottis. The fault in this piece of work was that the method of tearing out the spinal accessory nerve was too violent. It disturbed the neighbouring roots of the vagus as well. The vocal chords are in fact controlled not by the spinal accessory nerve, as Bernard thought, but by the vagus and his method of experimenting was not delicate enough to distinguish between the two.

The role of the pancreas in digestion

In 1846 Bernard observed that the urine of some rabbits which had been brought to him from the market was clear. This was unusual because the urine of herbivores is normally cloudy, but these animals had not been fed for some time and were therefore temporarily in the same state as carnivorous animals. They were in fact living off their own tissues. Bernard found that a very hungry rabbit would eat cooked lean meat and that under these conditions its system resembled that of a carnivore and its urine was clear. These observations led him into a physiological investigation which in turn resulted in his discoveries about the chemical properties and functions of the pancreatic juice.

After killing rabbits which had been fed on cooked lean meat, Bernard dissected them and found that the lacteals at a distance of 30-50 cms. below the pylorus (i.e. the lower opening of the stomach), were full of milky chyle. Now he knew that the lacteals in the dog were usually full of milky chyle *immediately* below the pylorus and he wondered what caused this difference. He then saw that in the rabbit the opening from the pancreatic duct is 30-50 cms. lower down in the duodenum than in the dog and he deduced from this that it must be the pancreatic juice which is responsible for making neutral fats absorbable. This was a new idea for the pancreatic juice had generally been regarded as almost identical to saliva in both composition and function. Indeed the pancreas was often called the abdominal salivary gland. Bernard was able to show that crushed pancreatic tissue when mixed with fats and kept at body temperature for a short time, caused the fats to decompose into fatty acids and glycerol. Thus the pancreatic cells clearly contained a substance which was capable of reacting with fats when brought into contact with them.

A second function of pancreatic juice was observed by Bernard in a sample which had been stored for some time out of contact with the air so as to prevent putrefaction. He found that when chlorine water was added to this sample a red coloration was produced. This reaction had already been described by Tiedmann and Gmelin in 1831. It is now known to be due to pancreatic digestion during which the proteolytic enzyme trypsin breaks down the available protein into its constituent amino-acids, including tryptophane. It is the latter substance which then produces the red coloration on oxidation with chlorine water.

The third property of pancreatic juice which Bernard described was its ability to convert starch into sugar. In this pancreatic juice does resemble saliva. He showed that all these reactions could be brought about both by pancreatic juice and by crushed pancreatic tissue, but he explained its action on the fats wrongly. Bernard thought that the process of emulsification was a physical process entirely and missed the action of the alkali in pancreatic juice which is necessary in order to bring about the dispersion of the fat droplets to form a suspension. Bernard was writing in 1850 before the colloid chemistry of Thomas Graham had been introduced and he can therefore be excused for misinterpreting this complex process.

In his animal experiments Bernard had taken pancreatic juice from no fewer than 34 dogs. He had found it to be a limpid, colourless fluid which became thick on agitation. It was alkaline in reaction, had no odour and a taste similar to that of blood serum. When heated it was found to coagulate in a manner similar to egg-white. In his attempts to obtain enough of this fluid for his experiments, Bernard had noticed that there was no secretion of pancreatic juice in the fasting animal; only when food was given and digestion began did the secretion appear and it seemed that the passage of acid chyme into the duodenum was the signal for the secretion of pancreatic juice to begin. Had Bernard followed this up he might have discovered the hormone secretin. He thought that a continuous flow of pancreatic fluid represented a pathological condition, but it is now thought that there *is* a small continuous flow which is greatly increased by the presence of food or secretin.

Bernard tried unsuccessfully to extirpate the pancreas from dogs. This would have produced a diabetic condition and in fact he did describe one case of a depancreatised dog in a profound state of emaciation, yet with a voracious appetite as might be expected in diabetes. The condition had been caused as a result of his attempts to produce a pancreatic fistula in this animal. After its death an autopsy showed that it had virtually no pancreas at all. Strangely enough, Bernard did not recognise the diabetic condition in this case, although he wondered whether the result might have been produced by the suppression of the pancreas.

Since he had not been successful in extirpating the pancreas completely by surgery, Bernard tried blocking its ducts with a form of wax which remained solid at body temperature. This caused undigested food to pass out of the body. He concluded from his experiments that pancreatic juice by itself acts on fats and carbohydrates, but not on proteins. The softening and gradual dissolution of lean meat in contact with pancreatic juice was regarded by Bernard as the early stages of putrefaction and he thought that proteins dissolve only on putrefying. He observed that pure pancreatic juice would not dissolve proteins completely, that when the food left the stomach it was first acted upon by the bile and only later by pancreatic juice and that proteins, having been in contact with bile were then completely dissolved by pancreatic juice. It appeared that pancreatic juice could only successfully attack proteins after they had been acted upon by the bile. The real explanation for Bernard's observations is that pure pancreatic juice contains little or no *active* trypsin. This enzyme is present but in an inactive form which requires the presence of enterokinase, a constituent of intestinal juice, to convert it into its active form.

The glycogenic function of the liver

Bernard's most significant discovery in physiological chemistry was concerned with the function of the liver in synthesising a carbohydrate from the constituents of the blood. He was not only able to demonstrate the capability of the liver in this respect, but went on to isolate this compound which he called 'glycogen'. Before 1850 when Bernard made his famous statement on this subject it was generally assumed that only plants possessed the power of synthesising complex substances from simpler ones, whilst animals on the other hand broke down the more complex substances by oxidation. If this view is accepted then all the parts of the animal body must be contained as such in its food, and indeed there were many who contended that this was the case. In France, for example, Dumas gave this concept the weight of his considerable authority as an organic chemist. It will be recalled that when Dumas and Boussingault published their famous essay on *The Chemical and Physiological Balance of Organic Nature* in 1841, the main theme was this difference between the synthetic activity of plants and the opposite, oxidative degradation of complex substances brought about by animals. Dumas even held that the butter of cows' milk all came from the grass which formed their food. Liebig protested at this and suggested that the fat in an animal was derived from sugar and starch by loss of oxygen.

The problem of the origin of fat in milk raised the parallel issue of the source of milk sugar. This too was thought by Dumas to come directly from plant food and Bernard accepted this view at first. Assuming that no sugar was to be found in the body except that taken in with the food he began with the intention of discovering where in the body the sugar was burned up. During some lectures on the digestion of starch given in 1846, Magendi had injected a solution of starch into the veins of a rabbit and within a few minutes had found not starch but *sugar* in the animal's blood. On another occasion he reported the presence of sugar in the blood of a dog fed on starchy food. These observations indicated that sugar is a normal constituent of healthy blood, although previously sugar had only been known to exist in the blood of diabetics.

In 1848 Bernard and Barreswill showed to the Academy of Sciences in Paris a sample of alcohol which they had obtained by fermenting sugar extracted from the liver of a dog. They were able to show that even when dogs were fed entirely on meat this sugar was still formed and could be detected in the liver. Bernard was puzzled as to the origin of this sugar in the liver of meat-fed dogs. He argued as a physiologist, that if the liver were capable of secreting sugar then it must be regarded as a *gland* and as such its functions must be controlled by the brain. This led him to make experiments in attempts to stimulate the liver into action by interference with the appropriate areas of the brain. He found that by wounding the floor of the fourth ventricle of the brain of a dog or a rabbit he could cause quantities of sugar to appear both in the blood and in the urine. This state he called 'artificial diabetes' and the way in which it was caused indicated that the nervous system did in fact control the secretion of sugar in the mammalian liver.

These experiments were reported in 1849 and in the same year Bernard performed his crucial experiment by which he was able to confirm that sugar is produced by the liver. In this experiment a dog was kept for some time on a diet from which both sugar and starch were entirely excluded. Then soon after a meal, whilst digestion was actively proceeding, the animal was killed by transecting the spinal cord just below the base of the skull. Three sets of veins were then rapidly ligated. The mesentery veins carrying blood from the intestines to the liver, the veins from the pancreas and spleen and the portal vein at a short distance below the liver. Above the ligature on the portal vein there would be blood which had passed through the liver and had run back into the portal vein. This was the only sample of blood which had passed through the liver and when the blood in each of the ligated veins was tested for sugar this was the only sample which showed its presence. From this it seemed clear that the blood entering the liver contained little or no sugar whilst that leaving this organ was loaded with large amounts.

One of the greatest difficulties faced by animal and physiological chemists has always been that of devising sufficiently delicate tests. In these experiments Bernard used the process of fermentation in order to detect the presence of sugar. This is not delicate enough to detect the minute quantities of sugar which are present in *all* blood, though it was good enough to indicate the very large increase in sugar content caused by passage of the blood through the liver. It was in 1850 that his famous paper appeared in the *Comptes Rendus* of the Paris Academy of Sciences. By this time Bernard had devised an improved method of obtaining unmixed samples of blood before and after they had passed through the liver. In the following year he received the prize for experimental physiology, but it was not until much later that he was able to isolate and identify the compound glycogen itself.

The first use of the term 'matière glycogène' appeared in 1855 when Bernard spoke of an internal secretion of the liver containing this hypothetical substance which gave rise to sugar. Later in this year he set out to estimate the relative amounts of sugar present in the livers of animals in various states of health and nutrition. Like a good chemist he carried out all his analyses in duplicate and on one occasion, pressed for time he made the first analysis immediately after the death of the animal but left the duplicate until the next day. He was surprised to find more sugar in the second analysis than in the first and thinking that perhaps he had made a mistake he repeated the procedure, only to find the same result. The sugar content of liver increased on standing. Bernard next attached the portal vein of a freshly extracted liver to the water tap and flushed it out for forty minutes until no detectable sugar remained in the washings. By this time—after 1850 in fact—he was using a more sensitive test for sugar based on the reduction of copper potassium tartrate (Fehling's test was introduced in 1849). The washed liver was then allowed to stand for twenty-four hours after which sugar was again found in the diffusate and in the substance of the liver itself. Bernard went on to extract a white compound from liver which gave a positive test for sugar when moistened. It is strange to record that he did not at this time call his extract 'glycogen', but in March 1857 he gave directions for the preparation of glycogen which are almost the same as the modern method. Two years later, in 1859, he was able to show that in embryonic life the placenta begins to produce glycogen before the foetal liver is functioning. Life therefore begins with a special organ to supplement the glycogenic function of the liver and since Bernard had also found that the fluids which bathe the foetus contain sugar he concluded that the foetus is in fact diabetic. He later made histological studies of the liver and other tissues. In the liver he attempted to discover which parts of the organ were responsible for the production of glycogen and which parts produced bile. In other tissues, such as epithelium and mucus tissues, he found glycogen but he could not detect this substance either in the nerves or in the bones.

Not all physiologists accepted Bernard's views about the glycogenic function of the liver. To many it seemed doubtful whether this organ could really produce sugar and in any case the idea that all such compounds must be derived from plant foods was still current even in 1855. Hence criticisms of Bernard's methods and conclusions were not lacking and an interesting point of scientific method arose from this. In 1855 Figuier published the results of experiments which he had made and which, he said, showed that blood *entering* the liver via the portal vein of a dog fed only on meat, contained sugar if the blood were taken between two and four hours after food. On the other hand blood in the hepatic vein *leaving* the liver did not contain sugar. Bernard vigorously attacked this criticism saying that he found it difficult to understand how a man might find sugar in the blood when it was not there yet fail to find it when it was present.

The Academy was asked to investigate the rival claims and a committee was set up consisting of Dumas, Pelouze and

Rayer. The matter resolved itself into the question whether or not there was sugar in the portal vein carrying blood *to* the liver from the intestine in meat-fed animals. The experiments made by this Committee conclusively supported Bernard. Sugar was not found in the blood of the portal vein of a dog fed on cooked meat, but blood from the hepatic vein collected simultaneously did not contain sugar.

Now although Bernard appeared to have been vindicated he was not, in fact, justified in his claim that there was no sugar in the portal vein. This whole issue depended upon the delicacy of the chemical tests used. All blood contains some sugar but whether or not it can be detected depends on the experimental technique of the analyst. The committee, like Bernard, had used the fermentation test, whereas Figuier had used a more sensitive chemical test depending upon the reduction of copper ions. Had the committee used this test their results would no doubt have supported those of Figuier. The method employed by Bernard was not capable of detecting sugar below a concentration of 80-100 mg. per cent, the normal fasting level of blood sugar, and in view of this he was only really justified in asserting that *his* tests failed to reveal sugar in the blood of the portal vein. Yet, aware as he was of the difficulties attending the detection of sugar in blood, Bernard insisted that his results were *conclusive.* Even in 1856 Chevreul had shown that there is sugar in all blood and rather more in arterial than in venous blood. From this it might have been argued that sugar is carried to the liver in the arteries, but Bernard did not consider this possibility—he had settled the point about the glycogenic function of the liver to his own satisfaction and he considered all criticism of this conclusion as groundless. Carbohydrate metabolism is noted for its complexity and even now there are many points which have not been finally settled.

The action of poisons

The third major area in which Bernard made discoveries about
the chemistry of the body was in his studies of the action of
poisons, notably carbon monoxide and curare, the arrow
poison. In 1846 Bernard poisoned a dog with carbon monoxide
and then performed an autopsy on it during which he found
that the blood had become a bright scarlet even in the veins.
The same result was later observed in rabbits, birds and frogs
and after he had succeeded Magendie as professor of medicine
at the Collège de France in 1855, he went on to follow up these
experiments. He shook up measured volumes of blood with
oxygen, carbon dioxide and carbon monoxide separately.
Although oxygen and carbon dioxide were both absorbed, it
appeared that carbon monoxide was not since its volume
remained unchanged. Bernard found however, that the gas left
after shaking carbon monoxide with blood contained some
oxygen and it therefore followed that carbon monoxide had
displaced oxygen from blood on a volume for volume basis.
He next shook up one sample of blood with carbon dioxide and
a second with carbon monoxide afterwards submitting each
sample to an atmosphere of oxygen. The sample saturated with
carbon dioxide was found to take up a large volume of oxygen,
but that which had been in contact with carbon monoxide
absorbed none. Hence it seemed that the blood saturated with
carbon monoxide could not function, the gaseous exchange had
been paralysed, the physiological phenomena of respiration
were arrested and death resulted from asphyxiation. In an
indirect way Bernard had demonstrated by these experiments
that the red corpuscles of the blood are involved in the process
of respiration.

Continuing his experiments, Bernard was able to show
that oxygen is chemically combined in the blood. He injected
a solution of pyrogallic acid which is known to absorb oxygen,
directly into the veins of a dog. The blood was not deprived

of its oxygen but when the blood containing this pyrogallic acid came into contact with free atmospheric oxygen in the lungs, the pulmonary tissues turned black although the arterial blood itself was fully oxygenated. Bernard also shook up arterial blood with pyrogallic acid out of contact with the air. The blood retained its colour and only when air was admitted to the apparatus did the blood turn black. From these results he concluded that no chemical reaction takes place between the oxygen in arterial blood and pyrogallic acid and that it followed that the oxygen was chemically combined in blood. Thus his study of the action of carbon monoxide as a poison had led him to two fundamental points about the chemistry of the respiratory function of the blood, viz., that the red blood cells are responsible for the respiratory function and that the oxygen absorbed during respiration is not carried loosely in solution but is chemically combined with the blood.

Bernard had first become familiar with curare, the arrow poison, in 1844, when he was most impressed with the quietness of the death which it causes. He found that the curare must enter the blood stream through a wound, however slight and not via the stomach. It attacked the nerves at the point at which they entered the muscles, causing paralysis, but it did not seem to affect sensation. The victim simply lost his muscular powers successively. Bernard's experiments on curare furnished proof of the independent excitability of the muscles.

In some other experiments on the action of poisons, Bernard injected separately into two different veins at some distance apart in the body of a dog, a salt of iron peroxide (i.e. a ferric salt) and yellow potassium prussiate (potassium ferrocyanide). These chemicals traversed the body in the blood stream but only reacted at points where they came into contact in an acid medium. Prussian blue was therefore deposited in the stomach and in the bladder—the normal condition of the blood is alkaline. Using the same technique in another experiment, Bernard injected separately amygdalin and its ferment emulsin, neither of which cause death alone but which react together to release hydrogen cyanide leading to rapid death. In this case the reaction would proceed in the alkaline medium of the blood and it seemed that this fluid provided a medium capable of controlling the chemistry of the body by permitting some reactions to occur whilst preventing others.

The 'Milieu Intérieur'

About 1855 Bernard put forward the suggestion that the blood in animals forms an internal environment within which their body chemistry was controlled. Because of its alkalinity the blood prevented certain reactions but it provided, 'the theatre for all those phenomena of fermentation and combustion which accompany the vital manifestations of cells ...' This internal environment rendered the animal to a large extent independent of its surroundings and all living organisms adjust themselves so as to maintain their internal conditions constant. Bernard suggested that all the vital mechanisms, no matter how varied they seemed to be had the same ultimate objective of preserving the internal environment intact.

This concept, now called homeostasis, was developed more fully in his *Experimental Medicine*. The cells of living organisms are bathed by fluids constituting the internal environment. These fluids derive their constituents from the blood and life is maintained only so long as the composition of the fluids remains constant within narrow limits. Deviations in any direction from this constant composition arouse tendencies within the living organism towards restoring the balance. Now, this great principle, fundamental to physiology, was as much a prophecy as a deduction from experiments. It was from his complete familiarity with physiology in all its aspects that Bernard was able to propose the generalisation which has formed the basis for some of the most fruitful investigations of the present century. For Bernard himself the concept of the internal environment was much more than a mere philosophical speculation—it was to become an experimental tool by means of which physiologists might eventually gain control over the vital functions.

'If it happens some day that by virtue of patience and hard work, physiology does definitely become established as a science, then we shall be able by modification of the internal environment, i.e. the blood, to exercise our will on all this world of elementary organisms composing our body; when we know the laws which control their diverse relationships, we shall be able to regulate and modify to our taste vital manifestations.'

Chapter 10
Physiological Chemistry and Medicine again

IT WILL HAVE BECOME quite clear by now that animal chemistry flourished amongst physiologists and doctors. The applications of animal chemistry to medicine were manifest to some, but by no means all, physicians and those who saw the value of animal chemistry made bold attempts to base both diagnosis and treatment upon chemical principles. However, most medical men were influenced to a greater or lesser extent by the doctrines of vitalism and even the most mechanically-minded were not inclined to dispense lightly with the presumed influences of vital force in the complex chemistry of life.

Golding Bird (1814-1854), whose work in animal electricity has already been mentioned, recognised the importance of animal chemistry for the physician. Like many other physicians of the nineteenth century he was puzzled at the high incidence of the urinary stone and he made a detailed chemical study of the problem without coming to any striking conclusions. He employed Liebig's chemical 'equations' in his attempts to account for the formation of uric and oxalic acids, urea, carbonic acid and other excretory products, but he was not satisfied that it really represented the mechanisms of vital chemistry. Liebig had suggested that the vital force acted to shake free the atoms from their combinations in the tissues and that they then recombined together to form the excretory products. Golding Bird felt that it was necessary to ask *why* the atoms recombined as they did. He tried to supply an answer from his knowledge of the behaviour of the elements during electrolysis. Each atom of matter was endowed with an electric charge giving it a polarity and the capacity to combine with other atoms which was most marked when the atoms were in the free or 'nascent' state. Bird accepted the concept of vital force as a useful convention but considered that ultimately all the reactions of vital chemistry would be found to occur according to the recognised laws of chemistry. Yet he was led to think that the animal organism was somehow more than a mere chemical system. He was not convinced, for example, that Liebig's wholly chemical account of the origins of animal heat was adequate. Bird thought that electric currents in the nerves played some part in maintaining the body temperature and the animal functions generally. Thus, for Bird, the animal body was more than a mere chemical laboratory as Liebig had taught.

Animal Chemistry in the hospital

Perhaps it was necessary to have studied with Liebig in person to become fired with his unswerving belief in the efficacy of chemistry as the controlling force in the body. Certainly Henry Bence Jones was so stimulated by his brief contact with the great German master that he spent the greater part of his career in efforts to apply Liebig's ideas successfully to medicine. We have already commented on Bence Jones' first attempts to consider the common diseases of gravel, calculus and gout in the light of Liebig's theories of animal chemistry, but this was only the beginning of a long series of experiments on the changing composition of the urine in health and disease, carried out with the co-operation of patients in the wards of St George's Hospital in London. Bence Jones began with a number of tests on the proportions of earthy and alkaline phosphates excreted under controlled conditions of diet and exercise. There were found to be wide variations from time to time during each twenty-four hour period. After food there was always a noticeable increase in the quantity of earthy phosphates produced, whilst exercise caused a corresponding increase in the alkaline phosphates. Bence Jones gave his patients two meals a day, in some cases restricting them to vegetable food (bread with wine or tea) only, sometimes providing only animal food (meat with distilled water), but it appeared from the results that the nature of the food had little or no effect on the quantities of earthy phosphates excreted. Since the proportions of alkaline phosphates increased as a result of exercise, it seemed likely that these compounds were formed as a result of the metamorphosis of muscular tissues as suggested by Liebig, but this conclusion was not entirely borne out by the results obtained with patients suffering from nervous disorders such as St Vitus's Dance in which violent muscular exertion occurs. Even when the muscular contractions were so intense as to endanger the patient's life, there was no marked increase in the amount of phosphate excreted, although an increase in the *sulphate* content of the urine was noted. Bence Jones had also found an increase of sulphates—as well as phosphates—in cases of inflammation of the brain. The sulphur was thought to come from albuminous matter and the phosphorus from cerebral matter in such cases.

Bence Jones also investigated the changing acidity of the urine with great care. He found that during each twenty-four hour period the acidity of the urine 'ebbs and flows' in sympathy with the changing acidity of the stomach. The acidity of the urine was found to be at its peak just before food at times when the acidity of the stomach was at its lowest. Then, during digestion when the acidity of the stomach was at its greatest the urine became alkaline. Convinced that these variations were interconnected, Bence Jones was prepared to accept Prout's view that the digestive system as a whole worked on galvanic lines, the stomach and kidneys providing the positive pole, the liver the negative and sodium chloride in the blood acting as the electrolyte. Such a system would result in the secretion of hydrochloric acid in the stomach and soda in the bile and it would be possible to use the variations of acidity in the urine as evidence about the general state of the digestive system. The decrease in acidity of the urine after food was most marked and longest sustained after animal food alone; with vegetable food only the urine was generally rather more acid and since the nature of urinary deposits was known to depend largely on the acidity, it seemed clear that these results would be of value to the physician in correlating any deposit formed with the diet of the patient.

Some of the acidity in the urine after vegetable food was found to be due to sulphuric acid and in the course of his tests Bence Jones gave his patients doses of elementary sulphur and of dilute sulphuric acid. In both cases increases in the acidity of the urine were noted, but by far the greatest increase was caused by potassium sulphate. Turning to other potassium salts, Bence Jones gave doses of potassium tartrate, but this rendered the urine alkaline. When he tried ammonium tartrate he found, to his surprise, that this salt had no effect on the acidity of the urine and ammonium carbonate caused an increase in *acidity*. This led him to surmise that the ammonium ion was being oxidised to form nitric acid and on careful investigation he found that this was the case. Other ammonium salts were also found to yield increases in the nitrates present in the urine as were urea and certain alkaloids (e.g. caffeine) in large doses. Thus it was shown that some of the nitrogen in the body could be converted into nitrates by oxidation. Animal metabolism involved, therefore, not only the oxidation of carbon and hydrogen but also that of phosphorus, sulphur and certain compounds of nitrogen—oxy-salts of the last three elements being found in the urine.

Bence Jones' experiments carried out between 1845 and 1851 are interesting as attempts to study gross chemical changes in living human subjects. Through his connection with St George's Hospital Bence Jones was able to investigate the effects of diet, exercise and medicines on his patients in carefully controlled conditions. He applied Liebig's theories of oxidation to explain all his observations, but details of the causes were generally lacking and Bence Jones realised that the real knowledge of animal chemistry available in the 1850's was quite limited. Claude Bernard had described the procedures used by Liebig and his followers as an attempt to discover what was going on inside the house by noting what went in at the door and what came out by the chimney and in his later work Bence Jones turned to physical methods in order to try to trace more precisely the fate of certain elements *within* the body. This led to the introduction of the first biochemical tracers and Bence Jones proposed the concept of a chemical circulation within the body by means of which the action of medicines on the tissues might be explained. Before turning to this, however, we shall briefly consider some physiological work on respiration and diet carried out by Bence Jones' contemporary Edward Smith, in the interests of prison reform.

Smith's Respiration Experiments

Smith, whose work appeared between 1857 and 1861, began by considering the various methods which had been used for the experimental study of respiration. Prout and others had counted the number of inspirations in a given time, determined the proportions of carbon dioxide in a sample of expired air and then calculated the total quantity of carbon dioxide expired in the whole period. Scharling about 1843 had carried out some experiments on respiration in which he had seated a man in a box for one-and-a-half hours during which he had estimated the total expiration from the lungs and skin by absorption with caustic potash, but the very size of the vessel used in these experiments led to inaccuracy. In another method a steady current of air was drawn for thirty minutes through a face mask, the volume of which was equal to a single inspiration.

The subject in such an experiment was able to breathe normally and the total carbon dioxide evolved was measured, but since only a proportion of the air entered the lungs it was not possible to determine the ratio of carbon dioxide formed to total atmospheric oxygen. Smith considered these methods with a comparative table of results designed to show their inconsistencies. His own procedure was more advanced in technique and was the first really systematic attempt to correlate respiratory and metabolic processes in the human subject during muscular exertion.

The spirometer apparatus used by Smith consisted of a face mask made from sheet lead with inlet and outlet tubes of brass and having diameters similar to that of the trachea. The inspired air was measured in a dry gasometer whilst the expired air was passed first through concentrated sulphuric acid and then over caustic potash solution. The resistance to breathing in this apparatus was low and it could therefore be used for long periods, even during sleep. Smith also used his spirometer with prisoners who were undergoing punishment on the treadmill and in fact one of his objectives was to collect experimental evidence that this practice, then common in British gaols, was deleterious to the health of the prisoners. He found that one grain of carbon dioxide was produced for each 56.3 cu. in. of air inspired whilst resting, but this rose to one grain for 44.1 cu. in. when walking at two m.p.h. and one grain for every 39.7 cu. in. of inspired air when walking at three m.p.h. His results also showed that the evolution of carbon dioxide was directly connected with the intake of food as well as with muscular exertion. Thus in normal conditions the quantity of carbon dioxide evolved between one and two hours after food was about twice that evolved immediately before food. Good food and sound sleep were found to be accompanied by a rise in the evolution of carbon dioxide, but a rise in the temperature of the surroundings caused a fall in the quantity of carbon dioxide expired.

Smith also tried to distinguish between the effects caused by different kinds of food on the evolution of carbon dioxide in respiration. He measured the respiration rate and proportion of carbon dioxide evolved whilst sitting quietly, then ate specific foods and repeated the measurements every twelve to fifteen minutes until the effects had worn off. He found that foods containing starch, sugar and gluten were long lasting respiratory 'excitants', increasing the output of carbon dioxide over long periods of time. Milk, too, was found to cause an increase in the quantity of carbon dioxide evolved, but fats did not. Alcohol in general depressed the evolution of carbon dioxide but tea and coffee were found to be powerful respiratory excitants.

Despite meticulous care to produce exactly comparable conditions for his experiments, however, Smith found his results to be very variable. He concluded that the increase in evolution of carbon dioxide which accompanies muscular exertion was derived from carbohydrates and fats, but even when the quantity of carbon dioxide was increased five or six times by exertion, the excretion of nitrogen (as urea and uric acid) remained constant. Thus, increased repiration resulted solely in more rapid oxidation of non-nitrogenous matter. However the lack of consistency in the results serves further to underline the difficulty of making accurate quantitative experiments on the natural functions of the living body.

The Chemical Circulation

Bence Jones' animal chemistry appeared in a book published in 1850. This contained a summary of the results of his experiments on the composition of the urine and his conclusions about animal metabolism. It was criticised for the same faults as Liebig's work was said to show—too great a willingness to speculate on the basis of too little experimental evidence. A new edition of this book became necessary by the mid 1860's and, despite the earlier criticisms Bence Jones felt that the time had become ripe to attempt a fresh and yet wider synthesis which would cover the whole range of inter-relationships between animal chemistry and medicine.

The new book appeared in 1867 with the title, *Lectures on some of the applications of Chemistry and Mechanics to Pathology and Therapeutics.* In it Bence Jones considered the body as a combined chemical and 'mechanical' system. The law of conservation of energy had by then become generally accepted and Helmholtz had shown that energy is conserved within the animal system also. It had, of course, long been recognised that animal functions are fundamentally chemical and it now seemed clear that the energy of the animal system was derived from the chemical affinities between inspired oxygen and the elements of the food and tissues. These oxidative reactions which occur in the body under the influence of the nerves were the source of animal heat, muscular energy, nervous energy and in total of life itself. Since they were known to occur in every *cell* of the body, it followed that there must be a diffusion of chemicals to all parts of the animal system. The general distribution of chemicals, aided by the circulation of the blood, could be explained in terms of Graham's work on diffusion and colloids, together with the process of osmosis.

From these ideas Bence Jones proposed the concept of a *Chemical Circulation* in the body by means of which the chemical substances present in the cells were distributed throughout the living organism. The chemical circulation was one of the few properties common to both animals and plants. By its means all living things could derive their chemical constituents from their surroundings. In animals the chemical circulation was to be regarded as of equal importance with the mechanical circulation of the blood, with which of course it was connected.

It was not only the natural constituents of the body which were considered to circulate, but also added chemicals such as medicines must also undergo a similar diffusion into the tissues where they could act to correct chemical errors causing disease. In order to judge the relative effectiveness of chemical medicines it was clearly essential to determine just how far and how rapidly this diffusion occurred and for this purpose some form of 'tracer' was needed. Bence Jones investigated this problem using lithium salts which could be detected in minute amounts by means of spectrum analysis. 'It occurred to me,' he wrote, 'that both in animals and plants the spectrum analysis ought to determine with certainty where diffusing substances go to; how long they stay in the textures and how quickly they cease to appear in the excretions.'

Early biochemical tracers

The first experiments were made in collaboration with August Dupré, lecturer in chemistry and toxicology at the Westminster Hospital. Dupré had already applied spectrum analysis techniques to the determination of lithium and strontium in London drinking water. In the new physiological experiments, guinea pigs were given doses of lithium carbonate either by cutaneous injection or in their food; they were then killed after measured periods of time varying from a few minutes to several days. Some of the organs and other parts of the body were then incinerated and the ash so obtained was examined for the presence of lithium by spectrum analysis. Particular attention was paid in these tests to the crystalline lens of the eye since this was regarded as a part of the body which is furthest removed from the vascular system and consequently most dependent on the process of diffusion through the tissues. By determining the earliest time at which lithium was to be detected in the lens a measure of the *rate* of diffusion through the tissues could be obtained. On the other hand some indication of the persistence of lithium in the body of the guinea-pig

Antoine Laurent Lavoisier.
(1743 - 1794)

Plate 9

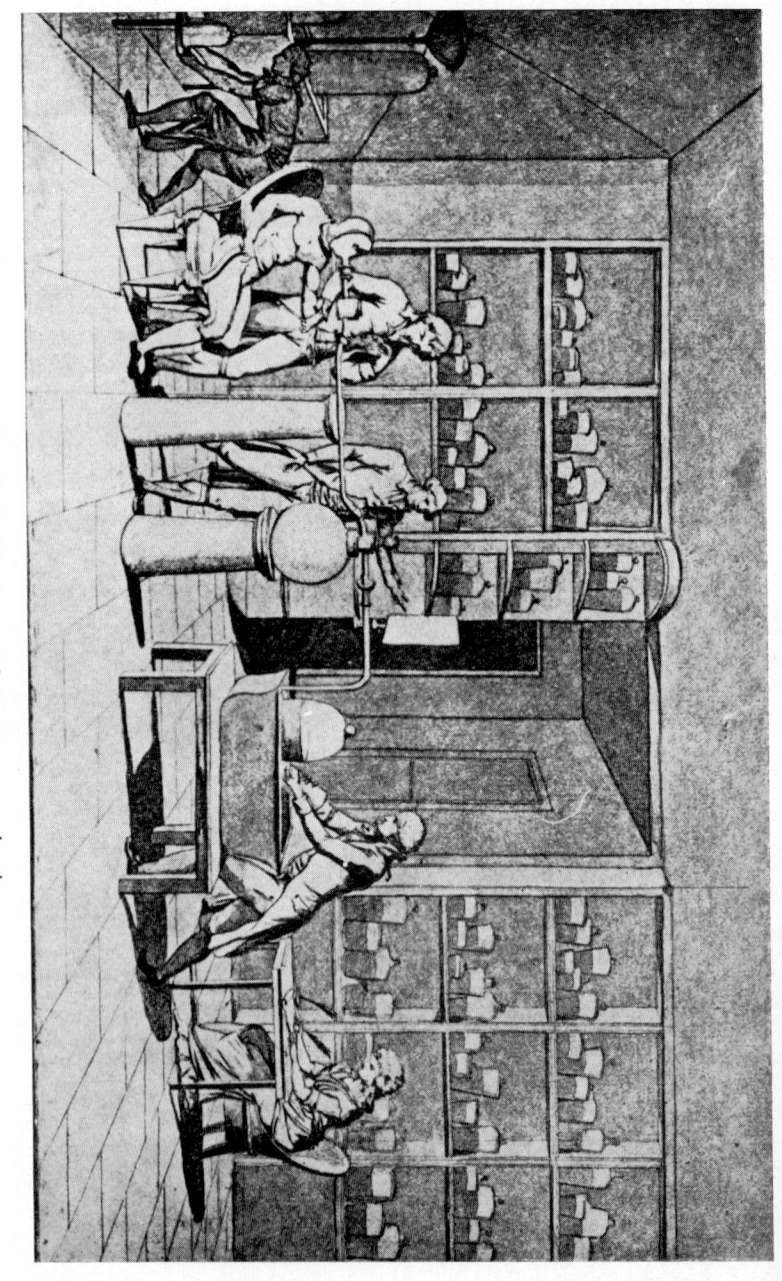

Plate 10. Lavoisier's experiments on respiration.
(Print supplied by the Open University)

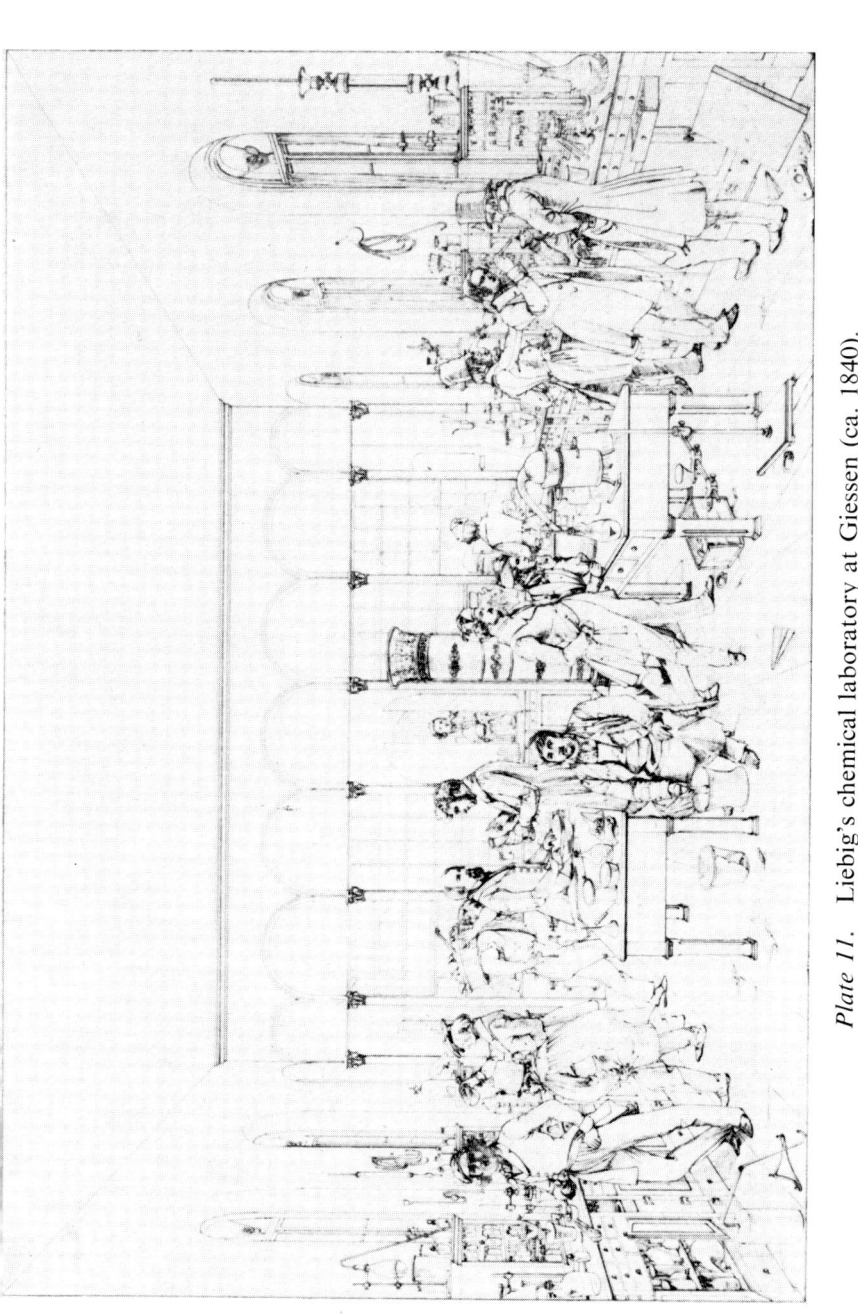

Plate 11. Liebig's chemical laboratory at Giessen (ca. 1840).
(Print supplied by the Open University)

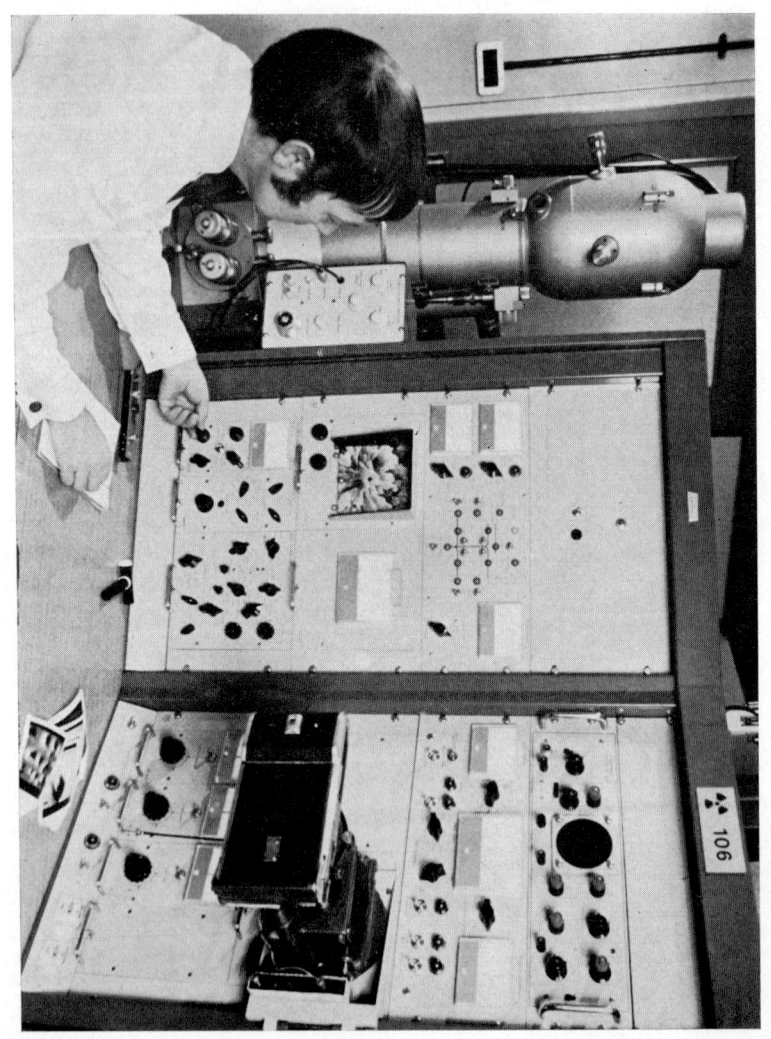

Plate 12.   A modern Electron Microscope.

could be obtained from an examination of the urine over long periods.

As in his earlier experiments, Bence Jones was able to apply these techniques to human subjects as well. He chose to test the rate of diffusion and persistence of lithium in the human body by applying his method to patients in St George's Hospital who were awaiting operations for cataract in which the lenses of the eyes were to be extracted. Thirteen such patients were asked to take doses of lithium carbonate at different intervals before their operations and then the lenses extracted from their eyes were examined for the presence of lithium by spectrum analysis as before. It was found that in the human subject, within three-and-a-half hours of the dose a trace of lithium could be detected in the lens and even after four *days* this element was still present. Urine tests on the same patients showed similar results.

This method seemed acceptable for the detection of certain metal salts in the tissues. Bence Jones therefore tried to extend it to compounds of other metals including rubidium, strontium, silver and even the highly toxic metal thallium, using guinea pigs as before. Although all the tests indicated that diffusion had occurred, the sensitivities of all these metals were much lower than that for lithium and the results of the tests were not encouraging. It appeared that the spectrum analysis technique could not provide a general method for the detection of minute amounts of elements in the tissues.

Many medicines contain in addition to the mineral salts substances such as the alkaloids and it seemed at least likely that the chemical circulation must extend to these as well. Some alkaloids were known to fluoresce in ultraviolet light and Bence Jones suggested that this property might provide a means of detecting alkaloids in animal tissues. He began in this case with quinine which fluoresces strongly and found that with aqueous solutions of quinine sulphate the delicacy of the fluorescence test for quinine was almost as great as that of the spectrum analysis test for lithium.

As before guinea pigs were used in the first experiments, and in this case extracts of their organs were made by boiling with dilute sulphuric acid, rendering the solution alkaline with potassium hydroxide and then extracting with ether. These extracts were then tested for fluorescence, but Bence Jones was surprised to find that whether or not the animal had been given quinine, the extracts showed fluorescence and this property was especially marked in the extract from the crystalline lens of the eye. The fact that animal *tissues* fluoresced in ultraviolet light was already well-known, but it had now been shown that a fluorescent *substance* could be extracted from them. Bence Jones went on to examine this animal extract and found that in optical and chemical properties it resembled quinine very closely—indeed he was unable to distinguish clearly between the two substances and he named the new extract 'animal quinoidine'. There are now known to be several substances with this fluorescent property in animal tissues and at least one of them, called thiochrome, is much more fluorescent than quinine itself.

When the animal had been dosed with quinine the fluorescence produced in the extract from its organs was found to be intensified, but by expressing the natural fluorescence in terms of a standard solution of quinine sulphate it was possible to obtain a quantitative measure of the intensity of fluorescence due to the alkaloid. Bence Jones traced the rate of diffusion of quinine into the tissues, determining both its amount and duration. He found that in the guinea-pig the alkaloid passed into all the vascular tissues within fifteen minutes, reached a maximum concentration in three hours and had entirely disappeared after three days. He was also able to make some tests on human subjects from which he gave an estimate of the earliest time after dosage that an alkaloid drug could be expected to act and some idea of the duration of its effects.

The ultimate objective of this work was to determine the

general chemical effects which medicines had in the body and thus to add to the precision of therapeutics. Bence Jones' results would no doubt be of value to the physician, but they are probably of greater importance to the development of bio-chemistry for the experimental applications of biochemical tracer techniques which they demonstrated. The modern bio-chemist has been enabled to develop and extend these tech-niques by the introduction of a wide range of radio-active isotopes produced by the irradiation of normal elements in the nuclear reactor (see Chapter 12).

Of course, Bence Jones' work was still concerned only with the overall changes of animal metabolism and all the complex details were left unexplained. In recognising this Bence Jones pointed out how difficult it was to express in detail even the apparently simple processes leading to the formation of carbon dioxide and water vapour in the body, suggesting that it must clearly be *much* more difficult to understand the manner in which blood, bone, muscle and nerve tissues were formed from the food. Liebig had held that whilst all such complex processes were fundamentally chemical, they nevertheless pro-ceeded under the influence of the vital force. Bence Jones, on the other hand, thought that all the chemical reactions of the body were purely chemical and allowed the concept of vital force merely as a general method of accounting for (but not of explaining), those natural functions for which no satisfactory chemical explanation was yet available. Thus, for Bence Jones there was no general vital force controlling all the chemistry of life, but the 'vital force component' was relegated to the ever-diminishing unknown factors which would ultimately be removed altogether. Although perhaps Bence Jones applied his simple chemical notions too directly to animal chemistry and life processes, we can see in his conception of the part to be played by physical and chemical techniques for investigating such processes the true pioneer of biochemistry. The road to the future study of the chemistry of life begins to open out in the work of Bence Jones as in that of other pupils and followers of Liebig.

The chemistry of the brain

We shall close this chapter with a brief account of some important contributions to physiological chemistry made by another pupil of Liebig and pioneer in the subject, J. L. W. Thudichum (1829-1901). Of German descent Thudichum emigrated to London in 1854 following his disappointment at the political discrimination against him which prevented him from obtaining a good medical post in Germany. His learning was very wide; it included a good knowledge of music, literature and the classics as well as chemistry and medicine. All through his life he continued to practise medicine whilst at the same time carrying out important and successful researches in physiological chemistry. His researches included a study of gallstones and another of the pigment urochrome found in the urine. He prepared haematoporphyrin from haemoglobin by the action of dilute sulphuric acid and he was the first to characterise the carotenoid pigments. His major work however was concerned with the chemical constitution of the substances of the brain, published between 1874 and 1882 in the Reports of the Medical Officer to the Privy Council and collected in 1884 into his great *Treatise on the chemical constitution of the brain*. In the course of this work he isolated kephalin and distinguished it from lecithin, a compound isolated from egg-yolk, bile, venous blood and other animal substances by Nicholas Golbey, a Parisian apothecary. These compounds are examples of the phospholipids, a class of substances first recognised by Vauquelin in 1812. Amongst them Thudichum discovered sphingomyelin and he also introduced the term *phosphatide* to describe a group of compounds for which he claimed a place of great importance in the chemistry of life. Thudichum was also well aware that these compounds are particularly abundant in the matter of the brain. His studies of the physical and chemical properties of the phosphatides led him to suggest a chemical explanation for several forms of mental disease. Thudichum also discovered an entirely new class of compounds in the brain which he called 'cerebrosides'. These were found to have a sugar-type structure, but they were also basic in character due to the presence of amino groups in their molecules. One such compound, phrenosine, was found by Thudichum to contain a new base which he isolated and called sphingosine $(C_{18}H_{37}O_2N)$. In phrenosine this base is condensed with a sugar which Thudichum called cerebrose, but which is now known to be identical with galactose, and with a fatty acid. The grey matter of the brain was also found to contain lactic acid already discovered in muscular tissue by Berzelius. Thudichum showed that the lactic acid in the brain is optically active and he also isolated another cerebroside in the brain which he called kerasin.

These discoveries are impressive enough in themselves, but it was the way that he applied them so as to construct a chemical account of the diseases affecting the brain, that Thudichum showed his real imaginative genius. In his studies of the properties of phosphatides he had found that they readily form an emulsion when mixed with water. The particles were too small to be seen singly but the solution exhibited optical properties and Thudichum regarded the emulsion as an intermediate state of matter between the fluid and the solid. He suggested that at the normal temperature of the body, some of the phosphatides might be in true solution but become colloidal at the higher temperatures occurring for example, in fever. Such changes he thought might be the cause of death in some cases of high fever and exposure to excessive heat. He also supposed that the so-called softening of the brain began by loss of the colloidal state and that the principal diseases of the brain and spine such as paralysis, acute and chronic mania, melancholy etc., resulted from such changes in the neuroplasm. In other cases, decomposition of the cerebrosides in the white matter of the brain gave rise to diseased conditions. Thus Thudichum made several important suggestions about the chemical nature and functions of the brain in disease which might well have revolutionised the treatment given to patients suffering from mental and nervous disorders. However his discoveries received far less than their due recognition, largely because he was unfortunate enough to offend established scientific opinion both in England and in Germany.

The protagon controversy

In 1864, Oskar Liebrich, professor of pharmacology in the University of Berlin, had announced the discovery of a substance in the brain which he called 'protagon' and which he claimed comprised the main part of cerebral matter. This substance was said to be present in all forms of cerebral matter and was common to the brains of all animals. Liebreich had given the formula $C_{116}H_{241}N_4PO_{22}$ for protagon and claimed to have decomposed it by hydrolysis. He had found it to contain a base which he called neurine, the platinum salt of which was given the formula $C_5H_{14}HCl_3Pt$. Glycerophosphoric and fatty acids were also obtained from protagon by hydrolysis with baryta water and at the same time the vinyl base $N(C_2H_3)$ $(CH_3)_2OH$ was formed. This compound had already been discovered by Hofmann who had synthesized it from trimethylamine and ethylene dibromide, hydrolysing the compound so formed with moist silver oxide. The name neurine was later reserved for this compound which is very poisonous and, like choline, is only known in solution. Liebreich's neurine obtained from protagon, was shown later to be identical with choline discovered in bile by Strecker and in 1866 the composition trimethyl hydroxyethyl ammonium hydroxide was proposed for it by Baeyer.

Liebriech's protagon was later thought to be a mixture of cerebrin and lecithin. The latter was first obtained in a pure state by Hoppe-Seyler in 1881 when he was professor of applied chemistry at Tübingen. Since it appeared that protagon could always be separated from brain substance Liebreich concluded that it was always present and that all the other compounds which could be isolated from the brain were derived from protagon itself. In fact protagon was a mythical compound, although its existence was strongly supported by influential physiological chemists following Hoppe-Seyler in Germany and Arthur Gamgee, professor of physiology at Owens College, Manchester, in England. Hoppe-Seyler who founded the first journal for physiological chemistry in 1877, may have been jealous of Thudichum's priority in the discovery of haematoporphyrin which he never fully acknowledged. Gamgee described experiments which purported to confirm the existence of protagon and bitterly attacked Thudichum's work. The controversy which arose led to a hardening of attitudes and the doctrine of protagon was upheld by the established authorities in physiological chemistry until the end of the nineteenth century.

Thudichum was undoubtedly very scathing in his criticism of those who supported the concept of protagon which he described as 'a most absurd fallacy'. He claimed that this idea, being incorrect, impeded the progress of science. The heat of the controversy was such that it resulted in the almost total lack of recognition both for Thudichum himself and for his work. Many unjust things were said and Thudichum was shunned by other physiological chemists who were his contemporaries. Yet throughout his life he stuck to his principles in the knowledge that he was right whatever others might say. The support of a few loyal friends and the strength of family ties enabled him to maintain his dignity in the face of the bitter hatred and opposition which his work on the chemistry of the brain had aroused. This controversy, like those in which Liebig had been concerned earlier, serves to remind us that scientists are after all human and are sometimes unable to rid themselves of those human prejudices which as scientists they must deplore.

The remarkable complexity of cerebral chemistry led Thudichum to suggest that perhaps mental diseases and insanity were caused by poisonous chemical impurities in the brain. He thought that poisons fermented within the body might cause insanity just as alcohol, fermented outside the body caused intoxication. If this were the case the time would come when such poisons would be isolated from the diseased brain and the poisons having been recognised, antidotes to them could be devised. Insanity too, would then be treated as a chemical disease and medicines could be prescribed for the condition as with physical diseases. In these far-sighted suggestions we can see Thudichum as the originator of the concept of biochemical lesions and autointoxications favoured by some psychiatrists today. He undoubtedly foresaw the possibility of treating such conditions in the light of the exact sciences. By a curious coincidence, lithium salts, Bence Jones' first 'tracers', are now commonly used in the treatment of certain mental disorders.

Chapter 11

Structural organic chemistry takes a hand

WITH THE APPEARANCE of Hoppe-Seyler's new Journal, *Zeitschrift für physiologische Chemie*, in 1877, biochemistry as a subject in its own right can be said to have come into existence. Pflüger, the editor of *Archiv für physiologie*, complained bitterly about the new publication. Physiological chemistry, he said, must be considered as part of a unified science of life; it belonged inseparably to physiology itself and had no place as an independent study. Hoppe-Seyler stood his ground, claiming that 'Biochemistry ... has grown into a science', and indeed by this time the broader outlines of the subject had been mapped out. The chemistry of fats, carbohydrates, proteins, purines, lipids and other important groups of organic compounds found in living matter had been described. The chemistry of digestion, respiration, nutrition and secretion had been studied and the existence of compounds such as enzymes and vitamins with specific biochemical functions was recognised, although few details about their chemical compositions were yet available. The rudiments of the subject were indeed available and it was time that the organic chemist began to apply himself to the study of the structures of some of these complex compounds.

Emil Fischer

The most prolific of the nineteenth century organic chemists whose work was important for biochemistry was the German chemist Emil Fischer (1852-1919). Kekulé's studies in structural organic chemistry had opened up a whole new approach to the subject which was nowhere more important than in the study of natural products. Chevreul, as we have seen, had long ago clarified the chemistry of the fats and it now fell to Fischer to devise techniques for working out the structures of the carbohydrates and the proteins, two classes of compounds which are fundamental to plants and animals. These studies required a high degree of chemical skill coupled with pheno-menal perseverance, fertile imagination and not a little luck.

Born at Euskirchen near Bonn, Fischer studied under Kekulé and von Baeyer. He became professor of chemistry successively at Erlangen (1882), Wurtzburg (1885) and Berlin (1892) and from the first his magnetic personality attracted research students to his laboratories. His early researches, which seemed to offer great promise, were concerned with colour chemistry, but he had determined that he would work on the chemistry of natural substances and in pursuing this ambition he went against the influence of von Baeyer who was working on the synthesis of indigotin.

When Fischer gave the Faraday Lecture at the Chemical Society in London in 1907, he chose as his title the phrase, 'Synthetical chemistry in relation to biology'. This admirably sums up his chemical work which must be regarded as having established the science of biochemistry on the firm foundation of structural organic chemistry. The variety and complexity of the saccharide molecules fabricated by plants from carbon dioxide and water during photosynthesis; the structures of poly-peptides and proteins and their construction from amino-acid molecules worked out from the results of chemical degrada-tions and attempts at synthesis; the power of building up or disrupting all of these complex natural substances which is exerted by the enzymes; these were some of the topics which Fischer investigated. He arranged his discoveries in coherent, intelligible sequences and in the process of studying the natural products he greatly enriched organic chemistry itself. His discoveries revealed a vast framework of structural organic chemistry within which those who followed him could work, filling in the details which his broad conceptual vision had passed over. In 1902 Fischer received a Nobel prize for his work and many of today's Nobel Laureates in biochemistry (Krebs is a good example), trace their inspiration back to him. In 1875 Fischer wrote a paper describing experiments on some derivatives of benzene diazonium salts. Amongst these he discussed dimethyl-, diethyl- and phenylhydrazines and his results led him to represent phenylhydrazine by the formula $C_6H_5.NH.NH_2$. Fischer investigated the reactions between this compound and certain aldehydes, including acetaldehyde, benzaldehyde and furfural. Thus, acetaldehyde condenses with phenylhydrazine to yield the compound acetaldehyde phenylhydrazone and the other aldehydes behave similarly.

Fischer did not, at this time, recognise the value of phenylhydrazine as a *general* reagent for the carbonyl group

$$CH_3.CHO + H_2N.NH.C_6H_5 = CH_3CH : N.NHC_6H_5 + H_2O$$

$>C:O$, and it was to be ten years before he returned to the subject to develop the full potential of this compound as a diagnostic agent for the sugars. In the interval he worked on rosaniline and the synthesis of caffeine; he was appointed professor of chemistry at Erlangen in 1882.

It was his discovery in the following year of a beautiful crystalline derivative by treating pyruvic acid $CH_3.CO.COOH$, with phenylhydrazine which led him to examine the reactions of this compound with carbonyl groups in general and the sugars in particular. He found that phenylhydrazine forms well-defined crystalline derivatives with the sugars, enabling them to be identified and separated from otherwise intractable syrups.

In 1886 chemists knew of only nine separate sugars including the aldohexoses glucose and galactose, the ketohexoses, fructose and sorbose, and arabinose, an aldopentose. Three hexobioses were known, sucrose, lactose and maltose, and one hexotriose, raffinose. Kiliani had given the accepted straight-chain formulae for glucose, galactose and fructose as pentahydroxy compounds and this was the sum total of knowledge about the sugars when Fischer began to work on them. In fact there are sixteen optical isomerides of the aldohexoses and eight possible racemic mixtures; Fischer was able to synthesize, formulate and determine the configurations for all but four of these twenty-four isomerides.

The key to the structural chemistry of the sugars lay in Fischer's use of phenylhydrazine. In 1884 he had found that this compound reacts with glucose to form glucosazone and that precisely the same product was obtained from fructose, whilst an isomeride was formed when galactose was used. Sucrose too, was found to yield glucosazone slowly, but maltose and lactose formed individual isomeric osazones. These reactions were found to occur with all the aldoses and ketoses, whilst the sugar acids yielded phenylhydrazides. All the derivatives were crystalline and quite distinctive, providing a ready means of identifying individual sugars. Also since they were crystalline compounds they could readily be purified and manipulated chemically.

Fischer himself said that his attention was directed to the sugars by his discovery in 1887 that the product obtained from the reaction of acrolein dibromide (CHBr.CBr.CH:O) with baryta consisted of two sugars isomeric with each other and with glucose. These two sugars were called α- and β-acrose. In 1913 it was shown that α-acrose is dl fructose whilst β-acrose is dl sorbose. Fischer had therefore synthesised naturally occurring sugars. They were optically inactive and he set out to resolve them by treating α-acrosazone with dilute hydrochloric acid by means of which it was hydrolysed to a tetrahydroxy ketonic aldehyde (glucosone). By the incomplete reduction of this compound using zinc dust, the aldehyde group could be hydrogenised before the ketone group and the process yielded fructose. Fischer applied this type of procedure to α-acrose and followed it by Pasteur's methods of resolving optical isomers. He was able to obtain sugars identical with d glucose, d fructose and d mannose. In 1890 he discovered that when any monobasic sugar acid was heated with pyridine or quinoline, the configuration of the carbon atom next to the carboxyl group became inverted or 'epimerised'. Using this process followed by reduction he was able to convert l mannonic acid into l glucose, d galactonic acid into d talose and d gulonic acid into d idose, respectively. He also obtained the aldopentoses l ribose from l arabinic acid and l lyxose from l xylonic acid. Fischer also made wide use of the cyanohydrin reaction discovered by Kiliani in 1881 and developed by him over the next five years. This reaction was useful for identifying the position of the aldehyde or ketone group in a sugar. Fischer said of Kiliani's research that it was 'the greatest advance in the investigation of the sugar group in the last decade.' Kiliani established the old formulae for glucose and fructose beyond doubt.

Fructose

$$CH_2OH \quad\quad CH_2OH \quad\quad CH_2OH \quad\quad CH_3$$

$$C{:}O \xrightarrow{\quad} C\!\!\diagdown^{OH}_{CN} \xrightarrow{\quad} C\!\!\diagdown^{OH}_{COOH} \xrightarrow{\quad} CH.COOH$$

$$(CHOH)_3 \;\; HCN \quad (CHOH)_3 \;\; hydrolysis \quad (CHOH)_3 \;\; reduction \quad (CH_2)_3$$

$$CH_2OH \quad\quad CH_2OH \quad\quad CH_2OH \quad\quad CH_3$$

methyl butyl acetic acid.

Glucose

$$CH{:}O \xrightarrow{\quad} CH\!\!\diagdown^{OH}_{CN} \xrightarrow{\quad} CH\!\!\diagdown^{OH}_{COOH} \xrightarrow{\quad} COOH$$

$$(CHOH)_4 \;\; HCN \quad (CHOH)_4 \;\; hydrolysis \quad (CHOH)_4 \;\; reduction \quad CH_2$$

$$CH_2OH \quad\quad CH_2OH \quad\quad CH_2OH \quad\quad (CH_2)_4$$

$$CH_3$$

n. heptylic acid.

Before 1891 it was necessary to *assume* a configuration for each sugar, but from that year onwards Fischer applied himself to the problem of assigning specific configurations to each sugar molecule. Applying Van't Hoff's concept of the tetrahedral carbon atom he showed that there must be sixteen possible stereoisomerides amongst the aldohexoses, eight possible aldopentoses and other related compounds. From their known chemical properties Fischer made a choice, naming the various compounds appropriately, and by 1896 he had been able to confirm his nomenclature for the monosaccharides.

He had discovered in 1892 that under the influence of dilute hydrochloric acid, glucose combines with methyl alcohol to form methyl glucoside. Since this substance showed no reaction with phenylhydrazine, it was considered not to contain an aldehyde group. It was stable towards alkalies but readily hydrolysed by dilute acids and it followed from this behaviour that the aldehyde group in the sugar should be considered to be combined internally with one of the hydroxy groups. This introduces asymmetry at the end carbon atom and there are therefore two optical isomers of methyl glucoside which Fischer labelled α and β.

β methyl glucoside α methyl glucoside

The methyl glucosides are the simplest amongst a group of compounds which also includes the naturally occurring substances amygdalin, indican and salicin. All yield glucose on hydrolysis. Fischer's studies of these compounds led him to elucidate the structures of some of the disaccharides and of glucose itself. He regarded the disaccharides as compounds like glucosides, between the hexoses into which the disaccharides are resolvable by the action of acids or enzymes. Maltose and lactose were given the general formula,

$$HO.CH_2.CH(OH).CH.(CHOH)_2.CH.OCH_2.(CHOH)_4.CH:O$$

$$\underline{\hspace{1.2cm} O \hspace{1.2cm}}$$

a————————————————b

in which the part (a-b) is glucose in the disaccharide maltose, and galactose in the disaccharide lactose. Fischer thought of the disaccharides as analogous to the synthetic glucosides in which the methyl group of methyl glucoside is replaced by a sugar molecule.

The two stereoisomers of methyl glucoside were found to behave differently towards enzymes such as maltase and emulsin. Thus, maltase found in yeast, would readily hydrolyse the α methyl glucoside to methyl alcohol and glucose but had no action on the β form, whereas emulsin, which is present in bitter almonds, had exactly the opposite effect. When, in 1894 Fischer found that emulsin hydrolyses lactose but has no effect on maltose, he concluded from this, together with the results of oxidation followed by hydrolysis, that lactose is glucose-β-galactoside and maltose is glucose-α-glucoside.

Amongst animal substances those containing nitrogen have always aroused interest because of the difficulties encountered in tracing the fate of this element in the body. Sugars in which one of the hydroxy groups was replaced by amino interested Fischer because they formed a connecting link between carbohydrates and amino-acids. In 1894, he and Tiemann investigated the structure of glucosamine which had been isolated from the chitin of lobster shells in 1878. The sugar-like compound which had been obtained by the action of nitrous acid on glucosamine was called chitose and was oxidised to chitonic acid by Fischer and Tiemann in 1894. Glucosamine was directly oxidised by the same means to chitamic acid. In 1903 Fischer was able to synthesise this compound from d arabinose and he showed that it could be reduced to d glucosamine. Chitose and chitamic acid were shown to be linked structurally with hydroxy-methyl pyromucic acid,

$$\begin{array}{ccc} CH & \!\!\!\!\!\!-\!\!\!\!\!\! & CH \\ \| & & \| \\ HO.CH_2.C & & C.COOH \\ & O & \end{array}$$

Chitonic acid, obtained from chitamic acid by the action of nitrous acid, is also related to this compound and it was concluded that chitose is derived from tetrahydro furane.

In the main Fischer's configurations for the sugars were given in terms of Kiliani's straight-chain formulae and it is now known that the sugars in fact contain oxide, or lactone ring structures. Yet Fischer's work had suggested the possibilities in this direction as well, for in 1893 with his discovery of the glucosides he had opened up the way to the recognition of a multitude of contingent isomerides. Thus he not only elaborated his own carbohydrate chemistry but he also laid the foundations for new developments in the subject which were to be continued in the work of W. N. Haworth.

The purines

From the opening of our study uric acid has been mentioned on several occasions and we have seen that the complex chemistry of its derivatives has always caused problems. Liebig and Wöhler had studied the relationships between the oxidation products of uric acid in terms of their molecular formulae in 1838, but it is to Fischer and his co-workers that we owe our knowledge of the molecular structure of uric acid and especially its relationship to the purines. In 1881, before he began work on the sugars, Fischer had been engaged in a study of caffeine. He was able to resolve this substance into methyl carbamide ($NH_2.CO.OCH_3$), and dimethyl alloxan. Theobromine, a related compound, on the other hand was resolved into methyl carbamide and methyl alloxan, showing that when caffeine was obtained by methylation of theobromine, it was to the alloxan ring that the new methyl group was attached. In other experiments Fischer prepared xanthine from guanine, oxidised xanthine to alloxan and carbamide, and by the reaction between the lead derivative of xanthine and methyl iodide he obtained theobromine. On the basis of such experiments he assigned the following formulae to xanthine, theobromine and caffeine,

NH.C:O.C.NH
| ‖   ⟩C.H
C:O.NH.C.N

Xanthine.

NH. C:O. C.N⟨CH₃
| ‖   ⟩CH
C:O.N.CH₃. C.N⟋

Theobromine.

N.CH₃. C:O.C.N.CH₃
| ‖   C.H
C:O.N.CH₃.C.N

Caffeine.

Fischer was even then interested in the relationships which might be found to exist between the molecular structures of these compounds and their physiological action. Between 1882 and 1900 Fischer and his students isolated about one hundred and thirty derivatives of uric acid and the purines.

Much of the fundamental work on the structures of uric acid and related compounds had been done by von Baeyer and others in the 1860's. The correct structural formula for uric acid was proposed by Ludwig Medicus in 1875, and in the same paper he gave formulae for a number of other purine derivatives including xanthine, theobromine and caffeine, pointing out their relationships with uric acid. Although his uric acid formula was firmly based on experimental evidence, the other formulae given by Medicus were not and they consequently received little attention.

$$\begin{array}{c} \text{NH.C:O.C.NH} \\ | \qquad \| \quad \rangle \text{C:O} \\ \text{C:O.NH.C.NH} \end{array}$$

Medicus' formula for uric acid, 1875.

In 1884 Fischer began with a study of methyl uric acid in the hope that he could establish the relationship between the methyl derivatives of xanthine and hypoxanthine. In the course of his work he discovered a number of new compounds which threw light on the structure of uric acid. By successive methylation he was able to ascertain the positions of the imide groups in uric acid and his results were found to confirm the formula given by Medicus. Fischer found that trimethyl uric acid is identical with hydroxy caffeine. As a result of such observations he was able to devise syntheses for caffeine and theophylline.

Whilst investigating methyl uric acid in 1884, Fischer had found that a mixture of phosphorus pentachloride with phosphorus oxychloride $(PCl_5 + POCl_3)$, replaced the oxygen atoms in the uric acid molecule. Two of the oxygen atoms were replaced readily and finally all three could be exchanged for chlorine to form trichloro methyl purine. This was in 1884, but thirteen years later in 1897 Fischer returned to these problems and was able to elucidate further the relationships between purine derivatives until in the following year he was able to isolate purine itself. He began by converting uric acid to 8 oxy-2:6-dichloropurine which could be reduced to 8-oxypurine, isomeric with hypoxanthine. This compound was also converted by ammonia into 6 amino- 8 oxy- 2 chloropurine which on reduction yielded 6 amino- 8 oxypurine, isomeric with guanine. Lastly oxidation of this compound gave 6:8

dioxypurine isomeric with xanthine. Also, by heating 8 oxy-2:6 dichloropurine with seventy parts of phosphorus oxychloride at 150°C., Fischer obtained trichloro purine, a base which yielded 7 and 8 methyl trichloropurines on methylation. From these results he had all the necessary information to synthesise hypoxanthine (6 oxypurine), xanthine (2:6 dioxypurine), adenine (6 aminopurine) and guanine (2 amino-6 oxypurine). He now felt ready to proceed to the isolation of purine itself. This he accomplished in 1898 using a two-stage process of reduction. Trichloropurine was first treated with hydrogen iodide and phosphorus tri-iodide at 0°C. and then with zinc and boiling water. The first stage yielded 2:6 di-iodopurine and the second gave purine itself.

2:6:8: trichloropurine   2:6 di-iodopurine   purine.

Apart from a few further minor researches on purine derivatives this completed Fischer's work on these compounds at this time. He then left the subject and did not return to it until 1914 when he was engaged on work concerned with the nucleotides—important chemical constituents of the living cell. In this work he used the silver derivatives of some of the oxypurines together with acetyl bromo glucose to form the d glucosides of theophylline, theobromine, adenine, hypoxanthine and guanine. All these compounds were found to be readily hydrolysed and were thus distinguishable from derivatives of glucosamine. Then by adding a cold mixture of phosphorous oxychloride and pyridine to a solution of theophylline d glucoside in pyridine, Fischer was able to prepare the first synthetic nucleotide, theophylline d glucoside phosphoric acid. Thus Fischer was able to link up his work on the sugars with that on the purines and to show that both groups of compounds are essential to all forms of life since nucleotides are universally found in all living cells. This work also represents the first steps leading to the study of the nucleo-proteins.

Amino acids, polypeptides and proteins

Proteins, like sugars, form a series of complex compounds the molecules of which are built up from smaller units, in this case the amino acids. Here then was another series of natural compounds admirably suited to Fischer's chemical genius. He began work on the amino acids in 1899 and he was ultimately able to show how they might be combined to form proteins. When Fischer entered this field there were nine known amino acids, three diamino acids and cystine, all of which could be obtained from proteins by hydrolysis or the action of enzymes. Of the known amino acids the dl forms of glycine, alanine, amino-valeric acid, leucine, aspartic acid, glutamic acid, phenylalanine and tyrosine had all been synthesised, but serine the ninth amino-acid, discovered by E. Cramer in silk-gum, was only synthesised in 1902. Erlenmeyer (jun.) obtained it from formic and hippuric esters; Fischer also synthesised serine (CH$_2$OH.CH(NH$_2$).COOH) in the same year.

The optical resolution of these compounds had proved to be difficult, but Fischer succeeded by benzoylating the amino group and then combining the resulting benzoyl-amino acid with an optical isomeride of strychnine or brucine. The d and l compounds then crystallised separately and the optical isomers of the amino acids could be obtained from them by hydrolysis. By this method optically active compounds could be obtained which were suitable for the synthesis of the polypeptides. Fischer also discovered in 1902 that by combining the hydroxy amino acids with β naphthalene sulphonyl chloride, he obtained well-defined, crystalline derivatives which could be used for the identification of the amino acids and even some simple polypeptides. Thus β naphthalene sulphonyl chloride assumed, although to a smaller extent, a role similar to that of phenylhydrazine among the sugars. It was an instrument for isolating readily soluble and otherwise elusive substances.

When a protein had been treated by hydrolysis most of the products remained as a syrup after the main constituents had crystallised. The amino acids in this syrupy mixture were very difficult to separate and identify, but Fischer facilitated this by esterifying them so that they would behave as aliphatic amines and he then went on to separate these by fractional distillation. He made an important advance by this technique for he provided the chemist with an analytical tool which could be used to separate the constituents of some of the most complex of all animal substances. The procedure is far from easy to carry out, but it is the only method which has been found successful in isolating the chemical components of complex protein molecules. It was later applied by Aberhalden, Fischer's brilliant pupil, to proteins such as edestin, elastin, fibrin, globin, gluten, keratin and the albumen in eggs, blood serum and milk. The component amino acids in amandin, excelsin, gliadin, hordein, phaseolin and zein were later ascertained by T. B. Osborne and his co-workers, whilst Fischer himself used the method to resolve the amino acids in the fibroin produced by spiders and silkworms. He remarked on the fact that despite the immense differences in their diets these two creatures produce a silk of almost identical composition. In 1907 Fischer examined the fibroin of the Madagascar spider and found the following proportions of amino acids present:

TABLE XII		
glycine	35.1%	
d alanine	23.4%	
l leucine	1.7%	
l tyrosine	8.2%	
proline	3.7%	
d glutamic acid	6.1%	
diamino acids	5.2%	(arginine)
ammonia	1.1%	
fatty acids	0.6%	

In the silk of the silk-worm, glutamic acid is absent but serine is present.

Fischer went on to study the possibility of synthesising the polypeptides and simple proteins. It had already been shown that amino acids would undergo autocondensation to form bimolecular anhydrides. Thus auto-condensation of the ethyl ester of glycine was thought to give diketopiperazine,

$$2 \; NH_2CH_2COOH \rightarrow \begin{array}{c} NH.CH_2.CO \\ | \qquad\qquad | \\ CO.CH_2.NH \end{array}$$

In 1900 Fischer obtained analogous products from the ethyl esters of α amino butyric, α amino *iso* caproic (leucine) and α amino *n* caproic acids. These he formulated as diketopiperazines and in the following year he went on hydrolyse the diketopiperazine from glycine gently so as not to detach the two glycine molecules from each other. The arrested hydrolysis of a cyclic anhydride thus began a series of experiments which by 1907 had led Fischer to the synthesis of a polypeptide containing eighteen amino acids, including fifteen glycine units and three leucine, thus:

$$NH_2.CH(C_4H_9).CO.(NH.CH_2.CO)_3.NH.CH(C_4H_9).CO.(NH.$$
$$CH_2.CO)_3.NH.CH(C_4H_9).CO(NH.CH_2CO)_8NHCH_2.COOH$$

This compound has a molecular weight of 1213 and Fischer calculated that there would be 816 possible optical isomerides for it.

Two other techniques were devised by Fischer for the synthesis of polypeptides. In one of these he treated an amino-acid or polypeptide with a chloro- or bromo-acyl chloride, e.g.

$$CH_2Cl.CO.Cl + NH_2.CH(CH_3).COOH \rightarrow CH_2Cl.CO.NH$$
$$CH(CH_3).COOH$$

In the other method which he discovered in 1904, it was found that phosphorus pentachloride would react with a substituted amino-acid to form a chloride which could be reacted with the esters of other amino-acids or polypeptides. This reaction could be used on d- and l- amino-acids and so could be applied to the synthesis of optically active polypeptides. By processes such as these Fischer was able to build up a polypeptide involving thirty amino-acid units, which he calculated to have 1.28×10^{27} possible isomerides. It seemed clear that the variety of possible protein structures would be almost infinite.

Enzymes, the biological catalysts

Within the living cell a complex series of reactions goes on and there is a continuous flow of matter in both directions through the cell wall. Nutrients enter and excretory products leave the cell by osmotic processes. Inside the cell the reactions which go on are catalysed by enzymes of which there are a very large number, each with its own specific function and all required because the reactions within the cell occur at low temperature. The enzymes guide the chemical reactions along metabolic pathways which lead to the synthesis of the organic constituents of living matter. The energy for these reactions is derived from oxidative breakdown processes which occur simultaneously. Thus the cell is an organised system in which one set of reactions is made to depend upon another. The detailed study of enzymes is relatively recent; in 1920, for example, only a dozen or so of these compounds were recognised, but now about 900 different enzymes are known of which nearly one quarter have been isolated in pure crystalline form. In fact this subject has become one of the most active growing points of biochemistry.

The first enzyme to be recognised was called diastase. It was a substance obtained from the aqueous extract of malt by

Payen and Persoz in 1833. They found that the extract was capable of converting starch into sugar and that it was inactivated by heating (a common property of enzymes). Pepsin, present in gastric juice was the next enzyme to be discovered. In 1834 Eberle had shown that digestion could be made to proceed in a test-tube if gastric juice were mixed with food matter and the mixture were kept at body temperature. It appeared that digestion must be regarded as a chemical process and was not dependent upon the presence of vital force. The presence of something besides hydrochloric acid in gastric juice has been recognised by Beaumont and in 1836 Schwann extracted an active principle from the wall of the stomach. This substance he called pepsin.

It is interesting to note that these discoveries came before Berzelius suggested the idea of catalysis in 1837. Berzelius had realised that catalysis could act in both inorganic and organic chemistry. He regarded it as a force quite different from ordinary chemical affinity; each catalyst seemed to have a specific function and the idea helped to gain an understanding of what is happening in living things.

'It gives us good cause to suppose that in living plants and animals thousands of catalytic processes are taking place between the tissues and the fluids, producing the multitude of dissimilar chemical compounds for whose formation ... we had not been able to think of any cause, but which in the future we shall probably find in the catalytic power of the organic tissue of which the organs of the living body consist.'

Whilst Berzelius merely *described* the process of catalysis, Liebig in 1839 had put forward a theory of enzyme action in fermentation processes in which he suggested that the enzyme (or ferment) transferred violent mechanical motion to the reactants by means of which they were shaken to pieces. Such changes were thought to occur at the outside surfaces of the cells but were not thought to extend inside. It was only in 1860 that Berthelot mascerated yeast cells and was able to extract from them and precipitate with alcohol, an enzyme which hydrolysed cane sugar, that the idea of enzymes occurring in cell contents arose. In 1878 Kühne introduced the name enzyme which he derived from the Greek words meaning *in yeast*. Then lastly, in 1897 Buchner obtained a cell-free extract from yeast which would bring about the whole fermentation process, indicating that the presence of living cells was not necessary for this process to occur.

The chemical nature of enzymes has caused some problems. In 1877 Moritz Traube suggested that they were substances allied to the proteins, capable of bringing about complex reactions without themselves changing. Later doubts began to grow about the protein nature of enzymes and some even thought that these bodies were not material at all but some kind of 'influence' still clinging to matter which had once been living. It was only after some enzymes had been isolated, purified and chemically examined that real credence was placed in the idea that enzymes are in fact proteins. This began about 1922 when Willstätter, working in Munich, purified saccharase and one or two other enzymes. He did not obtain pure products, though he made some progress in this direction. It was often thought that pure enzymes could not be prepared in any case, for in order to extract the enzyme it was necessary to crush the cells. This meant inevitably that the product was obtained from damaged cells and only studies on the constituents of cells which were intact could be significant. Purified enzymes must therefore be preparations which behave quite differently from the contents of the living cells. Here we can see once more the influence of vitalism—those who acted upon such beliefs would at once abandon the chemical search for the enzymes, considering it to be hopeless.

All the enzymes which have been purified during the past twenty-five years have turned out to be proteins. It is now known that they catalyse natural reactions by entering into the processes themselves and not merely by surface action as many of the inorganic catalysts do. Fischer, having worked on the structures of the carbohydrates and proteins, turned his attention to the study of enzymes of the saccharase type. He showed that each enzyme had a high degree of specificity for a particular substrate and in 1894 he suggested that the relationship between the enzyme and its substrate was similar to that between a key and the lock which it fits. Later work has shown this concept to be true to a degree which Fischer himself could hardly have suspected.

Biological oxidation

From the beginning of the 'modern' period of chemistry, starting with Lavoisier himself, it has been realised that the energy for the chemistry of life-processes comes from oxidation reactions within the body. What remained in doubt for so long was the mechanisms by which these biological oxidations were brought about. In 1840 C. F. Schönbein, professor of chemistry at Basel, discovered ozone, the allotropic form of oxygen. Molecular oxygen itself was known to be rather inert as an oxidising agent and Schönbein suggested that the conversion of oxygen to ozone was an essential first step in biological

$$3O_2 \rightleftharpoons 2O_3$$

oxidation. This idea became widely accepted and led to the nineteenth century 'ozone craze'. Thus in 1866-7 daily observations of the amount of ozone in the atmosphere were made in Paris and everywhere the relationship between this gas and states of health and disease formed a basis for speculation. Spas and health resorts, especially by the sea, claimed to derive their healthy climates from the ozone in their atmospheres. There is nothing in this correlation; ozone will not bring about biological oxidations and it is in fact toxic!

After the ozone theory there was a long period during which it was thought that the active agents in biological oxidations were peroxides. This idea was supported by the discovery in 1903 of the enzyme peroxidase which catalyses the oxidation of certain organic compounds by means of hydrogen peroxide. In this year A. Bach suggested on the basis of experiments with plant tissues, that the oxidation system of the cell could be separated into two parts. In the first of these oxygenase converted oxygen into a reactive organic peroxide whilst in the second stage peroxidase used this for the oxidation of other compounds. This theory dominated the subject of biological oxidation until about 1920, but it was later found that oxygenase was an enzyme, catechol oxidase, which was capable of oxidising catechol and other similar compounds by means of molecular oxygen. The products were not peroxides but quinones and peroxidase was not capable of using these as oxidising agents. Although it had been worked out from the results of experiments on plants, it was commonly thought that the oxygenase-peroxidase system also accounted for respiration in animal tissues. Catechol oxidase is however a typical plant enzyme.

Between 1920 and 1930 another theory of biological oxidation was proposed, based on experimental results derived from the respiratory systems in yeast, bacteria and animal tissues. On this theory the oxygen atom was first activated by combination with an iron atom in a special haemoprotein enzyme. This was called 'the respiratory ferment' by Otto Warburg, German physiologist and biochemist who won a Nobel prize in 1931 for his research into respiratory enzymes. The respiratory ferment was later (1925) identified with cytochrome oxidase.

About 1920 too there was another school of thought in which it was considered that it was the organic molecules which needed to be activated rather than oxygen itself. Most biological oxidations were seen to be dehydrogenations and a group of enzymes with this function—the dehydrogenases—were discovered. These are now known to be the most important factors in all biological oxidation systems and well over 150 different dehydrogenases are now known, each activating a different substance.

Thus we see that there were at this time two diametrically opposite systems which were proposed for biological oxidation. In one of these the basis was oxygen activation and in the other it was considered that hydrogen atoms in the organic compounds were activated. These two extremes were to be linked by the discovery of a number of intermediate steps brought about by the cytochromes discovered by Keilin in 1925, using spectroscopic techniques. It was found that the alternate oxidation by one system and reduction by another could be followed as it occurred in these haemoproteins, by means of the spectroscope. By observing the rate at which such changes occurred Keilin was able to follow the course of respiratory changes. In some cases there were as many as five different cytochromes acting consecutively, each reducing the next in line and the last reducing oxygen. The last cytochrome was then identified with the respiratory ferment. Many other intermediate stages have since been identified and it has been found that these intermediates are necessary to utilise all the energy available from the oxidation of proteins. This energy is stored in certain molecular configurations known as high energy bonds, the mechanisms of which are now being investigated.

The oxidation systems of the cell have been found to depend on the intracellular particles known as mitochondria—a granular part of the cell structure first isolated about 1948 by means of the differential high-speed centrifuge. Cell metabolism and the utilisation of energy within the cell was elucidated in the 1940's by Sir Hans Krebs and the cycle of changes which he postulated is still the subject of minute and intensive research. The complexity of enzyme action and biological oxidations in the living cell is so great that we can now begin to see why earlier animal chemists with their crude methods found their subject so difficult to decipher.

The vitamins

The vitamins are specific organic compounds which are required in the diet in very small amounts (in some cases as little as one part in 5×10^6). In the absence of these trace components of the diet, growth is retarded in young animals and pathological conditions develop in adults—avitaminoses. In man beri-beri, scurvy, rickets, pellagra, xerophthalmia and pernicious anaemia are known to be caused by vitamin deficiencies. Almost all our knowledge about the vitamins has been obtained in the present century and has resulted from two main sources, viz., the study of nutritional diseases in humans and by feeding purified diets to experimental animals so as to cause vitamin deficiency diseases. The vitamin has then been isolated from the food, its chemical composition and structure determined and then it has been synthesised. In each case in which this has been done successfully it has been found that the synthetic vitamin is completely equivalent to the natural substance in biological activity.

The first indications that there were unknown essential components in a healthy diet came from seafarers in the eighteenth century. Beri-beri and scurvy were common amongst sailors during long voyages and it was recognised that oranges and lemons would prevent scurvy even in the beginning of the century. These fruits were introduced into the ships of the East India Company. In 1753 Captain James Lind wrote a treatise on scurvy in which he recommended fresh fruits and green vegetables as a means of prevention and by 1804 a daily dose of lemon juice was made compulsory on all the ships of the British Navy.

During the siege of Paris in 1871 J. B. A. Dumas published

a paper on the constitution of blood and milk in which he described the plight of the people of Paris and the adverse effects which had resulted from the use of an artificial mixture intended to be a substitute for milk. In fact the scarcity of eggs and milk resulted in a very high mortality rate amongst infants and young children. It was thought that an emulsion of fat and albumen in a sweetened solution could be used as a substitute for milk, but although this mixture had something of the appearance of the natural product its effects on young children were disastrous and Dumas concluded that something essential to life was missing from the artificial milk. Although his statement of the case and of his own views regarding nutrition was clear enough, his suggestion that there must be minute quantities of unknown but essential ingredients in foods remained unnoticed for at least a decade.

Voit in 1881 suggested that if recognised nutrients were fed to animals in purified form there could be no doubt about their physiological value and the only reason why they might fail as nutrients would be their unpalatability rather than any lack of essential components. This was soon shown to be wrong however. In 1890, Christian Eijkman working in the Dutch East Indies discovered that he could cause beri-beri in chickens by feeding them exclusively on polished rice, the staple food of the human population amongst whom beri-beri was a common malady. It was found that an extract of the bran of rice would cure the disease and in the period between 1890 and 1897 Eijkman and his co-workers strove to extract and identify the active component of rice-bran. This is now known to be

$$
\begin{array}{c}
\text{CH}_3-\text{C}\underset{\underset{\text{N}----\text{CH}}{\|}}{\overset{\overset{\text{N}=\!=\!=\text{C.NH}_2\text{HCl}}{|}}{}}\text{C}-\text{CH}_2-\text{N}\overset{\overset{\text{CH}_3}{|}}{\underset{\underset{\text{Cl}\ \ \text{CH}----\text{S}}{}}{\overset{\text{C}=\!=\!=\text{C}-\text{CH}_2\text{CH}_2.\text{OH}}{}}}
\end{array}
$$

Thiamine

vitamin B_1, or *thiamine*, a water-soluble nitrogenous basic alcohol. It was for thiamine that Funk coined the name vitamine in 1912, although the compound was only isolated in 1926. Experiments were also made to test Voit's assertion by feeding synthetic diets to animals, but it was found that when quantities of purified fats, proteins, carbohydrates, salts and water which should have been adequate to produce good growth in an animal, were given as food, the animal did not thrive. Yet the addition to this purified diet of a quite insignificant quantity of milk made a remarkable difference and rendered the diet adequate. In the University of Utrecht about 1905, C. A. Pekelharing showed that mice did not thrive when fed on a cooked mixture of casein, egg albumen, rice flour, lard and salts, but when a very small amount of milk or whey was added to the diet the mice remained healthy. Pekelharing concluded that there was in milk an unknown substance which was necessary to enable the animals to assimilate the rest of their food. Without it, even when plenty of food was available, they died of starvation. Pekelharing's paper, important as it was, remained unknown and was only published in England in 1926. By this time Gowland Hopkins had already recognised the presence in food of traces of substances essential for health in an independent study published in 1912 and in the same year Funk had proposed the 'vitamine' theory. The hunt for these elusive compounds was on, although the real complexity of the situation with respect to the occurrence and functions of the vitamins was not realised for a long time.

Having mentioned Hopkins perhaps we should digress briefly to consider the importance of his work in establishing biochemistry as an academic discipline in its own right. Born in 1861, Frederick Gowland Hopkins' early scientific career began at Guy's Hospital where he became a medical student in 1888. Three years later, in 1891 he published a method for the detection of uric acid in the urine which was to remain standard practice for some years. His work on uric acid led to a study of pigments and in 1895 he published a paper on the colouring matters in the scales of butterflies' wings. This contained the first suggestion that excretory substances might be used for ornamentation and although his early findings were not supported by later investigations, Hopkins returned to the subject fifty years later and left some notes which merit further inquiry. In 1894, qualified as a doctor, Hopkins became an assistant in the physiology department at Guy's where he worked with E. H. Starling and Sir W. M. Bayliss. He realised early the need for much more precise knowledge about the chemical structures of protein molecules and studied the physico-chemical properties of these substances leading to the crystallisation of albumen from blood serum and egg white. In 1898 he was invited by Sir Michael Foster to join his illustrious school of physiology at Cambridge and he set about building an inspiring course of advanced study in chemical physiology, an aspect of the subject which was at that time neglected. In 1910, working with S. W. Cole, Hopkins isolated tryptophane from proteins and investigated the nature of the amino acids necessary for a mammalian diet which would provide both maintenance and growth. Already, in 1906, he had found that a diet containing only *purified* proteins, fats, salts and carbohydrates, in whatever proportions or amounts was not adequate for complete animal nutrition. Traces too small to contribute to energy values of unknown substances present in natural fresh foods—accessory food factors—were also needed. These discoveries formed the subject matter of a very important paper in this field, published in the *Journal of Physiology* in 1912, a paper which established Hopkins' reputation and for which he jointly received a Nobel prize with Eijkman of Holland in 1929.

In 1910 Hopkins had a year of illness as a result of which he was offered a praelectorship in physiological chemistry at Trinity College Cambridge. This meant that he had no formal obligations except his own researches and he embarked on studies of intermediate metabolism which he was to follow for the rest of his life. He and his students investigated the complex chains of linked chemical reactions, catalysed by intracellular enzymes which provides the physical and energy basis for the processes of life in general and cellular respiration in particular.

In 1921 he isolated and characterised the tripeptide glutathione which contains a reversibly oxidisable SH-group enabling it to act as a hydrogen carrier in biological oxidation processes and therefore a very important constituent for life processes in cells.

$$CH_2.SH$$
$$|$$
$$HC—NHCOCH_2.CH_2CH(NH_2)COOH$$
$$|$$
$$CO.NHCH_2COOH$$

Glutathione

Although he was forced to work in very restricted space Hopkins gathered round him over this period a school of young investigators. In 1914 the physiology department at Cambridge moved into a fine new building and Hopkins, now professor of biochemistry, was able to expand his work to fill its former quarters, but it was only in 1924 when the Sir William Dunn Institute of Biochemistry was founded that laboratory facilities for the subject became really adequate. A chair in biochemistry was endowed at that time, which Hopkins held until 1943. His aims were to explore intermediary metabolism and to establish biochemistry as a separate discipline concerned with this active chemistry of life processes and not *merely* with its fuels and end products. The most important contribution which he made to biochemistry was his insistence that biological problems could be solved in chemical terms as against the vitalistic ideas still in vogue when he began research.

To return to the study of vitamines; it was shown in 1915 that rats needed at least two separate chemical factors for growth. One of these, present in fatty foods was called *fat soluble A* and the other which occurs in non-fatty foods was named *water soluble B*. It was soon realised that the latter was identical with the anti beri-beri vitamine and, at the suggestion of J. C. Drummond in 1920, water soluble B was renamed *vitamin B*, the terminal 'e' being omitted because it was realised that not all of these substances were nitrogenous organic bases. At the same time fat soluble A was called vitamin A and some of its more important properties were noted. Lack of vitamin A led to lack of growth and increased susceptibility to infection of the respiratory system. The anti-scurvy factor was recognised to be different from either of these vitamins and was given the letter C, whilst vitamin D was found to be another fat soluble compound, distinct from vitamin A. In 1922 another factor needed for normal reproduction in the rat was described and named vitamin E. In this way the number of known vitamins steadily increased, and the vitamin alphabet grew, but it was soon found that vitamin B was itself a complex of several compounds.

In 1926 it was shown, as Funk had predicted, that pellagra was due to lack of a vitamin and that the particular compound involved was similar to the anti beri-beri factor, though more stable to heat. This was therefore called vitamin B_2 and the anti beri-beri vitamin (thiamine) was called vitamin B_1. It was

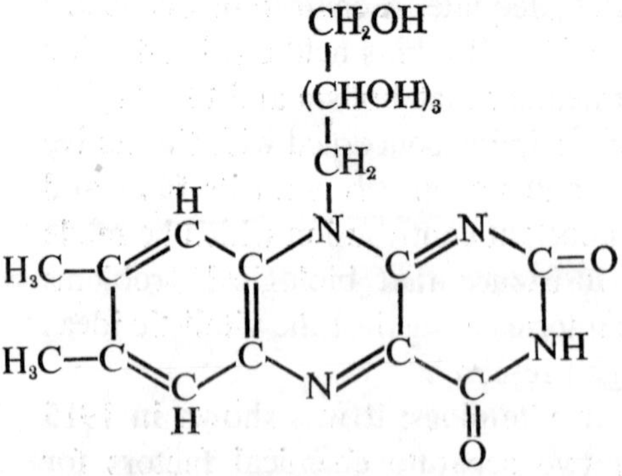

Riboflavin

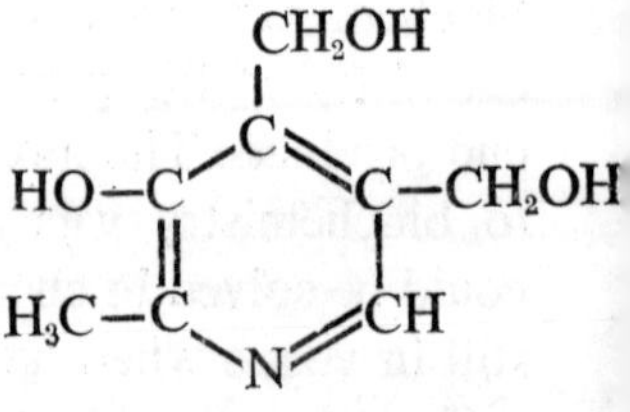

Pyridoxine

not long however, before it was realised that vitamin B_2 was itself a complex. The first component of vitamin B_2 to be isolated was riboflavin, the yellow pigment of milk, which in 1933 was shown to be the growth factor in rats. In the following year vitamin B_6 (pyridoxin pyridoxamine and pyridoxal), was identified, also by experiments on rats. Lack of this vitamin results in severe skin lesions. In 1937 the pellagra preventing factor proper—distinct from both riboflavin and vitamin B_6— was identified with nicotinic acid or nicotinamide.

Between 1928 and 1938 the better known vitamins A to E were all isolated in a pure state and their chemical constitutions were determined. All but vitamin D were synthesised and the artificial products were found to be identical in chemical and physiological properties with the natural products. Slowly other vitamins have been identified and synthesised or shown to be identical with already known chemical compounds. The details in this complex study of the vitamins are continually being filled in.

Hormones and the control of metabolism

Another branch of biochemistry which has received attention in the present century has also presented the organic chemist with many difficult problems. This is the study of the constituents of the secretions called hormones. These are substances secreted into the blood-stream by the endocrine glands. They arouse characteristic physiological responses in other body tissues, including the nervous structures, and so control metabolic rates in the body. The chemical and clinical aspects of hormone action have now become well-understood, but little is known of their reaction mechanisms. The hormones act upon the enzyme systems, either directly or indirectly by causing modifications in the genes responsible for enzyme synthesis. They are required to be present in extremely low concentrations in order to cause profound physiological effects and as a result

Vitamin	Sources	Effects of Deficiency
A (fat soluble)	Fish liver oils, butter, liver-fat. Also derived from carotene in carrots and green plants.	Xerophthalmia, affects eyes, skin, mouth, respiratory tract.
B_1 Thiamine (water soluble)	Pork, liver, whole grain	Beri-beri, brain, nerves, heart.
B_2 Riboflavin (water soluble)	Milk, egg-white, liver, green vegetables.	Lack of growth, changes in skin, mouth, eyes.
B_6 { Pyridoxine Pyridoxamine Pyridoxal	whole grain, yeast, egg-yolk, liver.	Red blood cells, brain, adrenals.
B_{12} Cobalamin	liver, meat.	Pernicious anaemia.
Niacin (nicotinic acid) (water soluble)	Yeast, wheat germ, meat.	Pellagra affecting gastro intestinal tract, skin and brain.
Folic acid (water soluble)	liver, green vegetables.	Anaemia.
Pantothenic acid (water soluble)	liver, kidneys, green vegetables, egg-yolk.	Adrenals, kidneys, skin, brain and spinal cord.
C Ascorbic acid (water soluble)	citrus fruits, green vegetables.	scurvy affecting bones, joints and mouth.
D (fat soluble)	fish oils.	rickets in bones and teeth.
E Tocopherols (fat soluble)	grains and vegetable oils.	muscles, red blood cells, liver, brain.
K (fat soluble)	green vegetables.	blood prothrombin.

TABLE XIII. *Some Common Vitamins*

both the hormones themselves and the chemical mechanisms by which they function are very difficult to detect. Moreover, the same hormone may act differently on more than one enzyme system.

Hormones fall into three main classes, viz., steroids; peptides and proteins; amino-acid derivatives, and although their presence in animal secretions has been suspected for a very long time the chemical complexity of these substances has meant that only in recent times has it been possible to isolate and synthesise some of them.

The term 'hormone' was introduced in 1905 by E. H. Starling, professor of physiology at University College, London, where he had worked on the secretion of pancreatic juice in conjunction with W. M. Bayliss. Their work had resulted in the discovery of the hormone secretin, a substance liberated by the duodenal mucosa. Secretin entered the blood stream and on reaching the pancreas stimulated that organ to produce pancreatic juice. It is now known that there are other hormones in the fluids secreted by the intestinal mucosa; pancreozymin also stimulates the flow of pancreatic juice whilst cholecystokinin causes the contraction of the gall bladder. These hormones are known to be of a protein or polypeptide nature, but their chemical structures and properties are yet to be investigated.

The pancreas itself is also known to secrete several hormones of which the best known is insulin, necessary for the metabolism of carbohydrates. The disease of diabetes mellitus has been known for a very long time, but its cause remained a mystery until 1889 when von Mering and Minkowski, both working in Strasbourg, removed the pancreas from a dog in order to investigate the part played by pancreatic juice in the digestion of fats. The animal showed all the symptoms of diabetes. Minkowski went on to remove the pancreas from other species of carnivores, showing in each case that diabetes resulted. He tried unsuccessfully to prepare extracts from the pancreas which would correct the disease, but it was only in 1901 that Opie described the damage to the islets of Langerhans in the interior tissues of the pancreas. This gave rise to the correct idea that a hormone secreted by these cells in the pancreas was responsible for the metabolism of carbohydrates, but it was not until 1922 that Banting and Best were able to prepare insulin in a form which made it available for therapeutic use. The mechanism of insulin is still far from clear. It is suggested that insulin activates the enzyme hexokinase which catalyses the phosphotylation of glucose to glucose-6-phosphate, a substance needed to release liver glucose into the blood. In addition to this insulin is also thought to aid the entry of glucose into tissue cells by reacting with cell membranes to produce physical changes leading to increased permeability for glucose.

The pancreas is known to produce another hormone, glucagon, which increases the rate at which glycogen is broken down and therefore raises the glucose level in the blood. Glucagon, which has been isolated in crystalline form, is found to be a polypeptide with 29 amino-acids in a sequence which has been determined.

The pituitary gland, situated below the brain, also secretes a number of hormones of protein or polypeptide nature which control the secretion of hormones by other glands. Several of

Cholesterol structure

these control the sexual functions whilst others govern the rate of growth and the chemical or physiological balance between the various parts of the animal body. Steroidal hormones are produced chiefly by the adrenal glands and the sex organs. They have been found to be related to the ring system of cholesterol and their chemical structures are all very similar, even small changes in structure producing profound physiological effects. The steroid hormones produced by the adrenal cortex control carbohydrate and mineral metabolism in the body. Some are concerned with the formation of glucose from proteins and enable the body to withstand stresses such as intense heat or cold, injury, and infection. Perhaps the best known amongst these substances is cortisone, which has been highly successful in the treatment of arthritis, allergies and ulcerative colitis.

Cortisone

The sex hormones, produced by the testes in the male, the ovaries in the female and the adrenal cortex in both sexes, are also steroids. These hormones control the sexual characteristics and in addition the male sex hormone, testosterone, stimulates the retention of nitrogen in the body by protein synthesis. The principal female hormones are the oestrogens and progesterone which is responsible for maintaining pregnancy. The chemical structures of all these compounds have been worked out and all are found to be remarkably similar.

TABLE XIV

Oestrogens

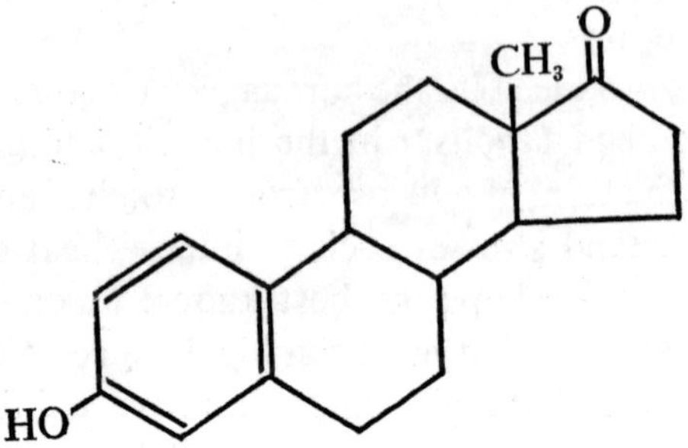

Oestrone

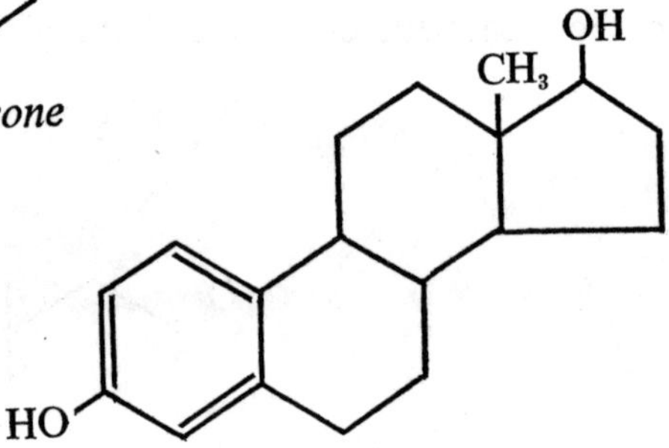

Oestradiol

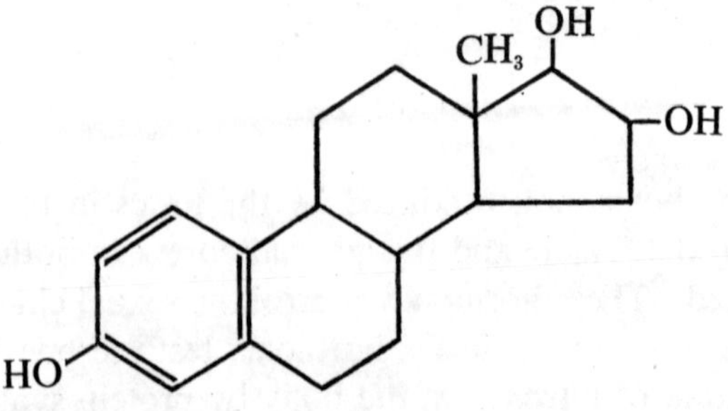

Oestriol

Testosterone

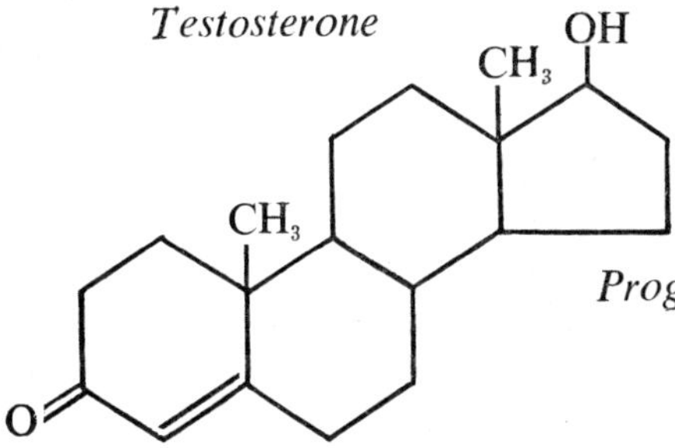
OH
CH3
CH3
O

Progesterone

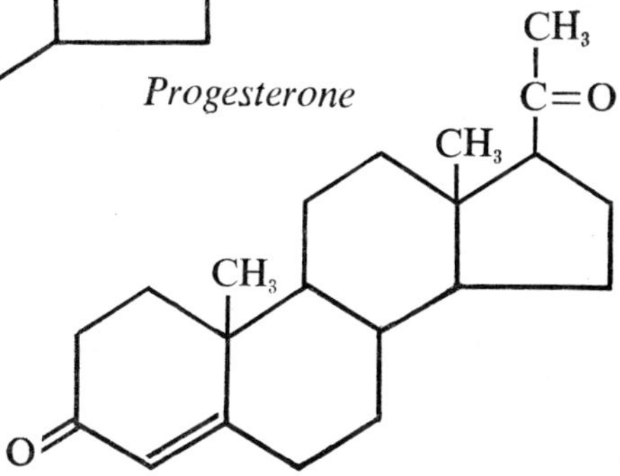
CH3
C=O
CH3
CH3
O

The inner tissues of the adrenal gland also secretes the hormone adrenalin, or epiphrine, which stimulates the sympathetic nervous system leading to constriction of the blood vessels and consequent rise in blood pressure. The pupils of the eyes dilate, hair stands on end and the eyeballs protrude. These phenomena are associated with emotional stress and it is clear that psychological factors are involved in the stimuli leading to the secretion of this hormone. Adrenalin was the first hormone to be isolated in crystalline form in 1901. It is a colourless solid, insoluble in water but soluble in both acids and alkalies. It is optically active. Adrenalin was first synthesised in 1904 starting from catechol.

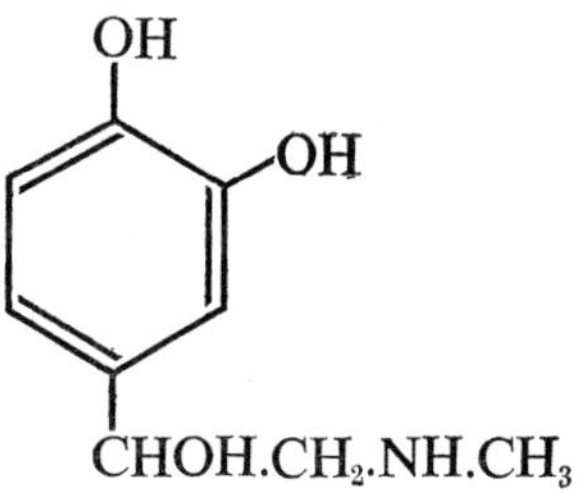
OH
OH
CHOH.CH2.NH.CH3

Adrenalin

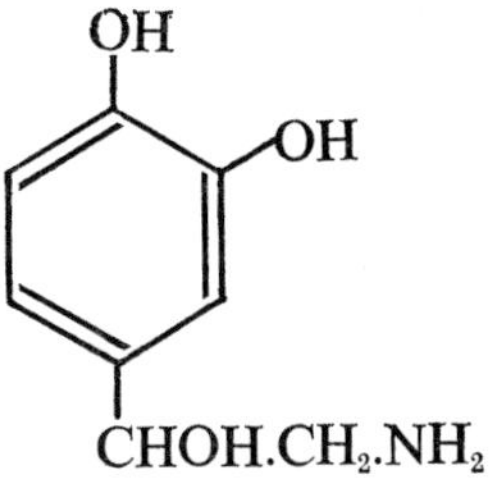
OH
OH
CHOH.CH2.NH2

nor-Adrenalin

Nor-adrenalin is also present in the adrenal medulla. It is laevorotatory and is thought to be mainly responsible for raising the blood-pressure.

These are amino-acid derivatives and similar hormones, although with more complicated structures, are produced in the thyroid gland in the neck. If this is underactive in children it results in retarded mental and physical development, whilst in adults decreased mental activity accompanied by swelling of the face and hands occurs. Lack of thyroid hormones is caused by iodine deficiency and in this case the gland increases in size producing the swelling in the neck known as goitre (Derbyshire neck). On the other hand excess of thyroid hormones leads to nervousness and the subject becomes underweight.

The main thyroid hormone is thyroxine. This compound is stored in the thyroid gland as a protein which is then hydrolysed by thyrotropin, a hormone produced by the pituitary. Thus

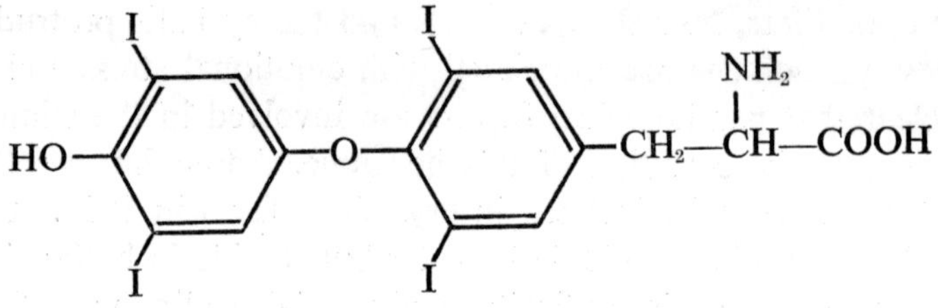

Thyroxine

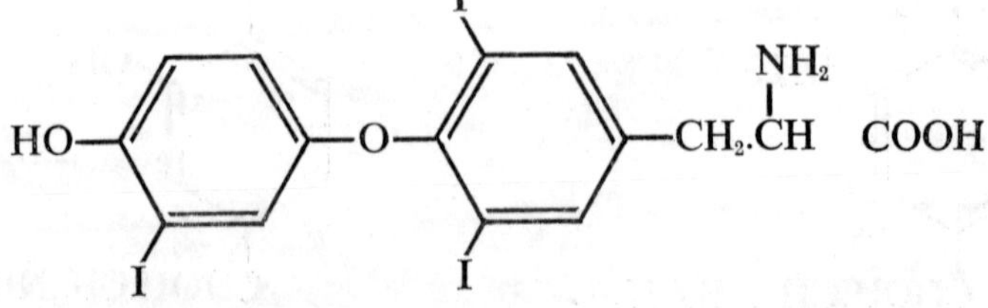

$3:5:3'$ tri-iodothyronine

one hormone calls another into action. There is also another compound which is present in very small amounts in the thyroid and in blood. This is $3:5:3'$ tri-iodothyronine and is between five and ten times more active than thyroxine itself. Proteins such as casein and serum albumen when treated with iodine also exhibit thyroid activity. Thyroxine has been isolated from hydrolysates of these iodinated proteins. Iodinated casein has been used to increase milk production in cows, egg production in hens and the growth-rate of young pigs. Thyroxine appears to increase the production of a wide variety of enzymes in the body. There is still much detailed work to be done.

The concept of macromolecules

Until about 1920 it was generally thought that organic particles of large molecular proportions, such as are found in colloids, must be made up of large numbers of relatively small molecules loosely bound together and capable of being separated by simple chemical methods such as hydrolysis. A false distinction was therefore made between colloid science and chemistry. In the first quarter of the present century however, evidence began to collect which indicated that very large molecules might well exist. Michaelis and Sørensen in their studies of enzymic proteins showed that these substances behave as polyelectrolytes and their solutions obey the phase rule. Svedberg found that haemoglobin gave a distinct band in the ultracentrifuge cell—a result indicating uniform particle size, whereas the aggregation of small molecules would give a range of particle sizes. Then in 1925-6 the molecular weight of haemoglobin was found to be about 68,000; it seemed that genuine 'macromolecules' might really exist.

This term had been coined by Hermann Staudinger (1881-1965), a German polymer chemist, in 1922. He defined the macromolecule as a *particle* in which the individual atoms are bound by normal valency forces. Staudinger was very articulate in explaining his theory and at the same time attacking the established ideas of aggregation. In 1925 he addressed the Zurich Chemical Society on the concept of the macromolecule and in the following year he spoke on the same subject at Düsseldorf. On both occasions he met with strong opposition; it was suggested that organic molecules containing more than forty carbon atoms could not exist and that if Staudinger thought that they did it was because he had failed to purify his substances sufficiently.

Some of the opposition arose because the aggregation theory harmonised well with the idea that colloidal particles were *supra* molecular in origin. Furthermore, X-ray crystallography, which had only recently been applied to cellulose, was thought to support the concept of small polymer molecules. Staudinger argued that chemical reactions such as methylation, acetylation, bromination, and hydrogenation of polymers should substantially reduce molecule size (and therefore molecular weight), on the older aggregation theory in which smaller molecules were thought to be held together by weak secondary forces which are attacked by these processes. In fact the molecular weights of the macromolecules are not substantially changed by these chemical reactions. Nevertheless, resistance to Staudinger's concept of the macromolecule continued and it was only in 1953 that his work was recognised by the award of the Nobel prize.

Staudinger had pointed out that the organic chemist in general worked at temperatures in the 100-200°C. range at which large molecules are destroyed. He felt that it was necessary to develop techniques of a more biological nature so as to treat the macromolecules at 'life temperatures'.

When, at last, the concept was accepted the myth that the secrets of life are hidden within *colloid* chemistry was exploded. Life functions were seen to be concerned with the large molecules themselves, although there remained the problem of the configuration of the macromolecular chains. In the early 1930's Astbury in England, Pauling in America and Wu in China independently suggested that in protein molecules the polypeptide chains were folded in precise forms which generate three-dimensional structures compatible with specific enzyme functions. On heating any protein these folded chains are straightened out and the protein is said to have been denatured. Confirmation for these ideas came from X-ray crystallography; Astbury showed that many fibrous proteins have folded structures and in 1935 he extended the same idea to the globular proteins. The convergence of the concept of the macromolecule with the early results of X-ray study of the natural fibres and single crystals of globular proteins can be said to be the essence of molecular biology and biological functions were thus correlated with the spatial arrangements of molecular constituents.

Chapter 12
Some techniques of modern biochemistry

IN TRACING THE RISE of those aspects of biochemistry which stemmed from animal chemistry we have seen the gradual change from a purely empirical applied science, owing its existence largely to the practical needs of physiology and medicine, to a complex theoretical study involving concepts developed in physical and organic chemistry. The animal chemists had to cope with ignorance, lack of techniques and the emotive approach of the vitalists. There was very little which could be said to be of value in building up *theories* of biochemistry before the closing years of the nineteenth century and the really useful physico-chemical techniques now employed by the biochemist have only become available in the present century. During the past fifty years or so, by the clever use of all the newly devised methods, biochemists have been able to construct a unified science with its own body of theories and laws from many inter-related strands.

Since about 1950 biochemistry has often been identified with molecular biology and it seems that it is at the molecular level that fresh advances can be expected to occur. Biochemistry, which is concerned with all types of living organisms, has as its ultimate goal the complete description of life at *all* levels including the molecular. Indeed, since all life processes originate from within the living cell it must be on the molecular level that biochemists should expect to make those discoveries which will enable them to extend the knowledge of life-processes. Growth, reproduction and heredity all fall within the orbit of the biochemist, since all occur at the molecular level within the living cell and all are closely related to the processes of cell metabolism. It is thought that all the properties exhibited by a living organism can be reduced to the properties of the individual cells. When we understand and can fully describe all the chemical changes which occur within the cell we shall have achieved the fullest description of 'life' possible by the use of the intellect alone. This is the aim of the biochemist but there is still a very long way to go before we can claim to understand life at all fully.

Physical techniques for the study of large molecules

From the purely chemical point of view most of the matter found inside living cells consists of sugars or other carbohydrates, fats or lipids and proteins. There are in addition, some nucleic acids and porphyrins—a class of red-pigmented compounds forming part of the active nucleus of compounds such as the chlorophylls, haemoglobin, cytochromes and the enzymes catalase and peroxidase. Each of these chemical groups

A typical proto-porphyrin

contains a wide variety of compounds and in many cases, especially when the molecular weights are low, the chemical structures have been worked out by the methods of organic chemistry. Determinations of the structures of such compounds as thyroxine and adrenalin and of some of the vitamins are good examples of such procedures. Nevertheless there are still large gaps in our knowledge, especially of those compounds whose molecular weights are high such as the complex proteins and the nucleic acids, compounds consisting of large numbers of smaller chemical units.

Large protein molecules generally contain several thousand amino-acid residues. It is fairly easy to determine the relative proportions of each of these by analysis, but the precise *order* in which the amino-acids in the protein chain is much more difficult to determine. Although there are only about two dozen different amino-acids involved in protein formation, there are no restrictions upon the order of their arrangement and consequently the number of possible isomers which can be formed from a thousand or more amino-acid residues is clearly immense. Yet this order is of the greatest importance, for it is from the amino-acid sequence that the specific properties of each protein are derived. In the 1950's Frederick Sanger, working at Cambridge, determined the complete sequence for the molecule of insulin (Mol. wt. 5700). Sanger's work opened the road to the determination of protein structures in general and it was an essential preliminary to the chemical synthesis of insulin for the treatment of diabetes. He was awarded a Nobel prize for his work in 1958. By 1963 the sequence for ribonuclease (Mol. wt. 12 700), had been worked out but when the same techniques are tried with proteins of molecular weights in the region of 10^5 or 10^6, problems of a different order are encountered. In such complex cases perhaps the computer can help.

One of the most successful physical techniques for investigating the structures of complex protein molecules is that of X-ray analysis. This has given information about the way in which peptide chains are coiled and folded in large protein molecules. X-ray crystallography, introduced in 1912 by Friedrich and Knipping, was studied by W. H. and W. L. Bragg, who used the technique to investigate the crystal structures of simple inorganic molecules. Since that time it has been developed into a powerful analytical tool by which crystal and molecular structures can be studied. In some cases (e.g. vitamin B_{12}) a large molecule can form a single crystal by itself and X-ray analysis then provides a means for determining the structure of the compound directly. The method was in fact applied to vitamin B_{12} in 1957 by Dorothy Hodgkin and her co-workers at Oxford. Proteins such as haemoglobin and myoglobin were also studied using this method by J. C. Kendrew and M. F. Perutz. The results of such X-ray analyses can be used to confirm and extend chemical analyses.

Earlier X-ray techniques involved the production of a photographic record from which determinations of the structure were made by visual estimation. The accuracy was not greater than 15-20 per cent, enough to solve the stereochemical problems but not precise enough to allow the determination of atomic thermal motions or types of chemical bonding. Since 1953 however methods of detecting diffracted X-rays by direct recording with Geiger, scintillation and proportional counters have been used. To these techniques is added the use of computed circuitry to synchronise movements within the crystal. All this is necessary for the investigation of proteins which are unstable. Rapidity coupled with accuracy is needed because of their instability and the large number of data to be collected.

The study of protein structures by X-ray diffraction methods continues to provide valuable information. During 1970 for example, Professor Hodgkin and her group at Oxford determined the three dimensional configuration of the insulin molecule, whilst a similar determination of adenosine triphosphate was carried out at Cambridge. At Purdue

University in America the three dimensional structure of crystalline lactate dehydrogenase, the largest enzyme molecule to be analysed so far, was determined. This important substance is one of the critical enzymes involved in the release of energy from glucose by anaerobic oxidation; it is present in most animal species. The Purdue group obtained their enzyme samples from the muscles of the dogfish and their results were so accurate that the exact point in the molecule at which it combines with co-enzyme diphosphopyridine nucleotide and the exact relationship between the four polypeptide chains in the structure of the molecule could be determined. In fact X-ray crystallographers think that they are well on the way to using their structural studies to explain in precise detail how enzymes function by observing the conformational changes which occur in these complex molecules during their chemical reactions.

Other techniques which have been used to investigate the proteins and other complex organic molecules include electrophoresis, optical measurements and the ultracentrifuge. We shall consider each of these briefly. Electrophoresis is in some respects similar to electrolysis except that the particles are colloidal in size and carry a variable number of electric charges adsorbed onto their surfaces. The colloidal particles move through the solution under the influence of the electric field and the speed of motion depends on the balance between particle size and the number of charges carried. The mass of the particles slows them down whilst an increase in the number of charges carried has the opposite effect. In some cases the particles are of microscopic size and can be observed directly under the microscope. When this is so the speed of the particles can be measured by direct observation, but more generally the motion of a moving boundary is observed. This may sometimes be coloured but is often seen as a change in the refractive index of the solution. It is possible to detect changes of refractive index as little as one part in 6×10^6. The concentrations of different components can be measured and substances which had been thought to be pure are sometimes shown to be mixtures. Thus it is now known that crystalline egg albumen and serum albumen each contain two components. Blood serums have also been analysed by electrophoresis and it has been shown that detectable changes occur in certain pathological conditions. Electrophoresis thus becomes a diagnostic tool.

The ultracentrifuge is used in the study of high polymers including proteins, nucleic acids, viruses and other macromolecules of biological origin. It can provide information on sedimentation rates and molecular weight distributions in polydisperse systems in which the particles in suspension are of various sizes. It can supply data on the frictional characteristics of large molecules from which the sizes and shapes of solute particles can be derived. The ultracentrifuge is also used for the separation of macromolecules and thus it becomes an analytical tool. The distribution of concentrations can be measured by optical methods whilst the centrifuge is in operation and optical absorption or radioactivity techniques can be applied to the residues after sedimentation.

The commonest optical technique for investigating molecular structure is that of spectrophotometry. For most routine purposes the spectrum of visible light is used. The sample is successively irradiated by narrow bands of wavelengths and the intensity of transmitted light for each band is measured by means of a photo-cell. Each compound yields a characteristic curve when intensity of transmitted light is plotted against wavelength. For chemical analysis the ultra violet region of the spectrum is often used. Organic compounds containing at least one unsaturated link absorb energy in the ultra violet. It is common to find that such compounds can be detected quantitatively and frequently only a minute sample is required because the absorption bands are intense. Thus, by using a long cell containing the sample, it is possible to detect as little as 0.00001 per cent of acetophenone dissolved in iso-octane and many compounds absorb much more intensely than this. For

the qualitative analysis of many organic compounds the infra-
red part of the spectrum can be used. In this case it is the rota-
tional energy and force in the chemical bonds which is observed
and since the spectrum produced is generally characteristic for
each compound, the method can be used for detecting the
presence of particular organic molecules.

When X-rays of sufficient energy are made to impinge upon
atoms and molecules, electrons can be expelled. The kinetic
energies and numbers of these electrons are characteristic for
given atoms, groups or chemical bonds and graphs drawn
with kinetic energy as abscissae and numbers of electrons
expelled as ordinates provide reliable evidence of the presence
of known chemical elements or systems. The method is sufficiently sensitive to detect the presence of single atoms or groups
in complex molecules. Thus the presence of sulphur in insulin
can be detected by this method as can the cobalt atom in
vitamin B_{12}.

It is by means of physical techniques such as these that the
modern biochemist has been able to determine the molecular
structures of the very complex compounds with which he has
to work. In no case has the problem been greater and its
solution more exciting than in the investigation of the nucleic
acids. Two types of these occur in every living cell. Deoxyribo-
nucleic acid (DNA) occurs in cell nuclei and is responsible for
transferring genetic information, whilst ribonucleic acid is
found chiefly in the cytoplasm and is the means employed in
the cell for protein synthesis. Both are very large molecules
containing nitrogen bases, phosphoric acid and a pentose sugar.
In DNA the pentose is β2-deoxy-D-ribose,

In RNA the pentose is β-D-ribose

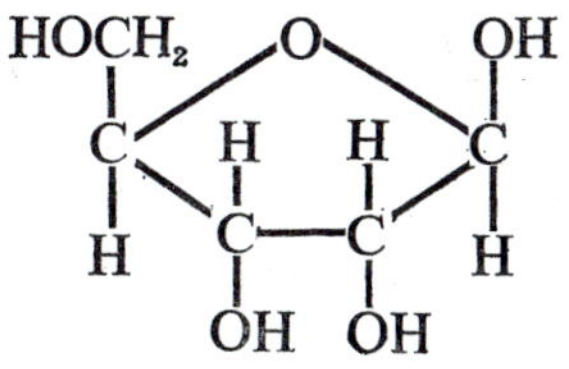

Both DNA and RNA contain long chains of alternating pentose and phosphate molecules with bases attached to the pentoses. These bases are of purine or pyrimidine origin. They include the purines adenine and guanine and the pyrimidines cytosine, thymine and uracil.

Table XV nitrogenous bases found in DNA and RNA

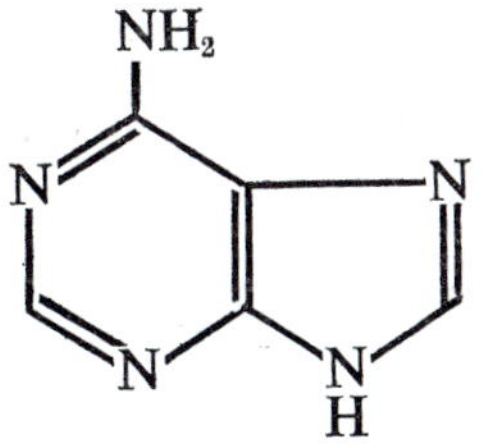

Adenine

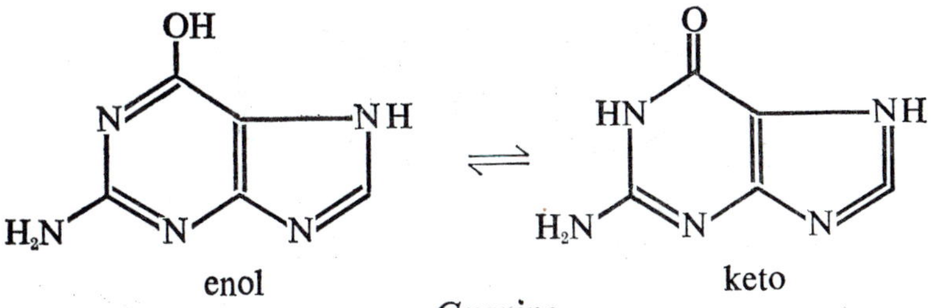

enol   keto

Guanine

Table XV continued

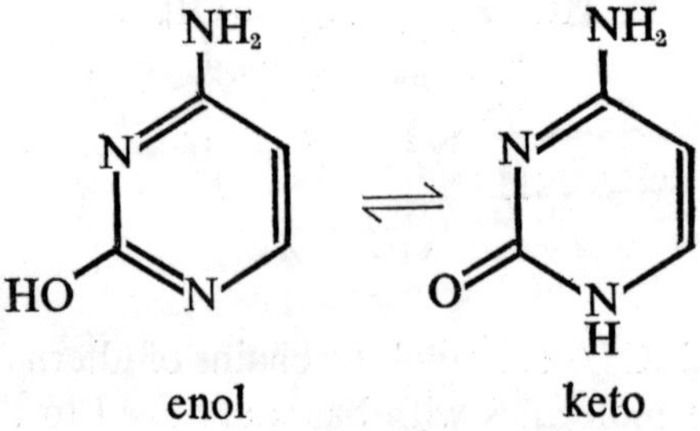

enol   keto

Cytosine

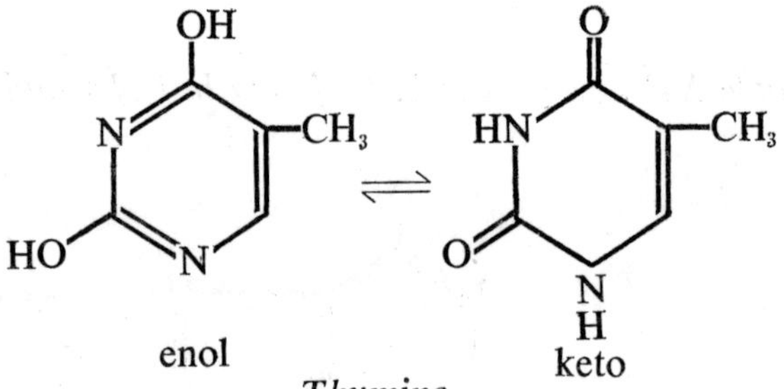

enol   keto

Thymine

enol   keto

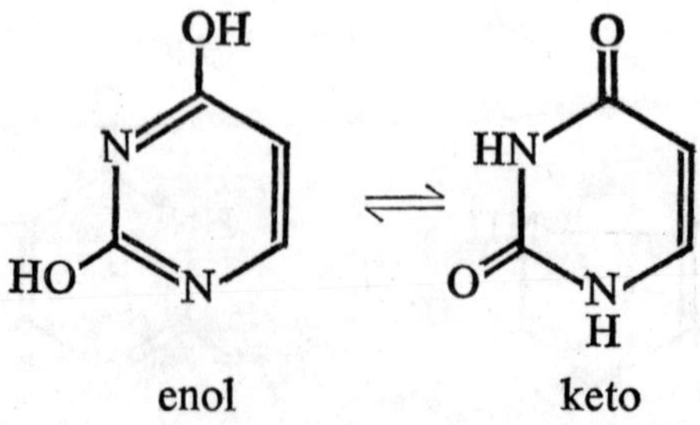

Uracil

In all cases where there is an oxygen atom in the position next to a heterocyclic nitrogen atom, keto-enol tautomerism can occur. By chemical analysis it was observed that in DNA adenine and thymine always occurred in equal proportions as did cytosine and guanine. X-ray analysis then revealed that the bond lengths between these pairs of bases were of the order of hydrogen bonds and starting from this information F. Crick and J. Watson in the 1950's proposed the well-known helical structure for DNA.

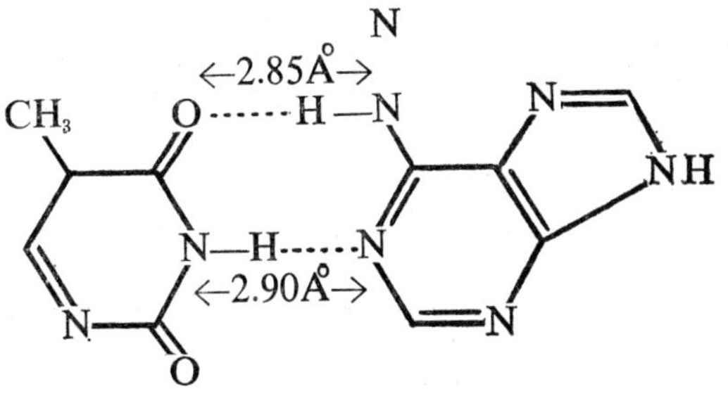

Thymine   Adenine

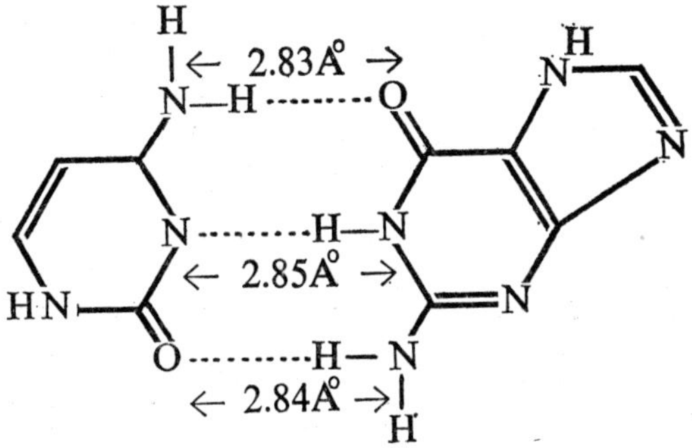

Cytosine   Guanine

Table XVI Base-pairs in DNA

It is now known that DNA carries the genetic code and that the long chains are made up of sequences of genes. The order in which the bases are arranged along the molecule of DNA determines this sequence which is then transferred to soluble (or transfer) RNA. The latter molecule wraps itself around a ribosome or ribonuclear protein particle in the cytoplasm of the cell and the arrangement of bases along the RNA determines the order in which amino acids are synthesised by the ribosome forming specific enzymatic proteins. The genetic information carried by the DNA molecule is thus transformed in the ribosomes into particular kinds of proteins, most of which by their enzyme functions are involved in the life processes of the organism. In this way the biochemist has come close to providing a completely chemical explanation of all the characteristics of living organisms and even human nature is included!

An important method of tracing the chemical changes which occur in animal functions became available with the discovery of radioactivity and since the introduction of atomic energy this method has been increasingly employed. The radioactivity of a radio-isotope is entirely independent of its chemical properties and its disintegration rate is dependent only on the number of radioactive atoms present. A radioactive atom may be incorporated into a chemical compound and then followed through a series of reactions. As little as 10^{-10} gm. can still be detected and many chemical changes can be traced with a high degree of precision. The usefulness of radio-isotopes as tracers in biology arises then from three properties:

  i. at the molecular level the physical and chemical behaviour of the radio-isotope is nearly identical with that of a stable isotope of the same element.
 ii. radio-isotopes can be detected in extremely small amounts.
iii. analysis for radio-isotopes can often be done without affecting the sample or system.

Radio-isotopes have been used to study steady-state systems, diffusion processes and metabolic pathways.

The preparation of compounds labelled with radio-isotopes is technically difficult. The elements labelled in a given compound must be in the same valency state. Purity is important because it is necessary to know that no other radioactive atoms are present in the chemical apart from the one specifically introduced to act as a tracer. When an exchange process is to be studied the compounds must be labelled in positions known to take part in the exchange, so that the transfer of radio-isotopes from one compound to another can be traced. On the other hand, if the compound itself is to be followed from one part of the body to another, care must be taken *not* to label it in an easily exchangeable position.

Radio-isotopes are detected in a variety of ways all of which depend upon the radiations emitted. Geiger and proportional counters are used for detecting α and β particles, but scintillation counters are needed for γ rays. In general the total radiation is observed, but in the technique known as γ ray spectroscopy the energy of each individual γ ray is recorded and by this method one radio-isotope can be detected in the presence of another because each produces γ rays of characteristic energy.

In physiology radio-isotopes are used in a wide range of permeability, absorption and distribution studies. The metabolism of calcium has been studied by injecting solutions containing Ca^{45} intravenously. It has been found that more than 80 per cent of the calcium leaves the circulation within five minutes, although the specific activity of the serum is still decreasing two months after the injection. These observations have been explained in terms of a balance between the calcium in different parts of the bones and the blood serum—the full story, as might be expected, is very complex.

Chromosomes can be labelled with tritium (H^3) which emits β particles. Tritium labelled thymidine is localised in DNA when cells are grown in a tritiated thymidine medium. These cells are then transferred to a non-radioactive medium for the development of the second generation. It has appeared that DNA does not take part in normal exchange processes and is synthesised only when the cell is preparing to divide. The rate and mechanism of DNA synthesis is at present under intensive study. The life-cycles of human red blood cells has been established at 120 days by observing Fe^{59} labelled cells. In addition the white cells and platelets have also been studied by radio-isotope labelling techniques. Some of these methods have applications in diagnosis and treatment of disease. Thus the uptake of I^{131} by the thyroid, the measurement of red-cell mass in the blood by Cr^{51} labelled red-cells and the absorption of vitamin B_{12} labelled with Co^{60} have all provided information of medical interest. A necessary part of this work is the study of radiation hazards, both as a result of radioactive fallout from atomic explosions and from the possible effects of radioactive pollution due to the increased use of the atomic reactor for the production of electric power.

In histological and cytological studies the method of auto-radiography is widely used. In this the specimen is made to produce an image on a photographic plate due to radiation emitted from radio-isotopes with which it has been treated. In general a somewhat blurred image is produced and in some cases the effect is so slight that the photographic plate needs to be examined with a microscope. When an electron micro-scope is used for this purpose single radioactive atoms can be detected; the technique is precise and delicate. Difficulty is sometimes encountered in making the specimen lie flat against the photographic plate. A whole plant or animal may be used though it is more common to prepare a thin section of the specimen. In some cases a thin film of the material is sprayed or spread on the surface of a microscope slide or sliver of newly cleaved mica. The method can also be used in conjunc-tion with paper chromatography, another extremely useful bio-chemical technique, and the radio-isotopes can then be traced on the resulting chromatogram. Most radio-isotopes produce radiations which penetrate the film to some depth and so produce an out-of-focus autoradiogram. Tritium however produces *low-energy* β-particles which are stopped at the surface of the film and so the points at which the film is affected are very close indeed to the radioactive sites in the specimen. This then provides a high-resolution autoradiogram which has become a powerful tool in detecting and measuring small amounts of radioactivity. Chromosomes labelled with tritium thymidine have been investigated by this method and it has been found that on replication each labelled chromosome yields one labelled and one unlabelled daughter chromosome. The whole process may require weeks or even months and when completed the developed film is examined by means of the electron microscope.

The magnification possible with the electron microscope is such that the instrument is capable of resolving macromolecules of biological interest. The material is sprayed onto the surface of freshly cleaved mica and a very thin layer of platinum is sprayed over the surface at an angle. Any projections above the surface then receive a heavier deposit of platinum on the side facing the metal-evaporating source and leave a 'shadow' with no metal layer on the far side. In this way particles of macromolecular size are picked out in relief. Globular molecules of diameters as small as 20-30 Å and linear structures such as the strands of nucleic acid of less than 10 Å thickness can be resolved. In fact the limits of resolution are set by the grain size of the shadow casting material. Thus the biochemist is equipped to examine the molecules of life.

Bioassay

An important aspect of the biochemist's work is the study of the precise effects of physiologically active chemicals such as proteins, vitamins, trace elements and therapeutic substances. These are tested under controlled conditions on suitable organisms. In all forms of bioassay the same basic principles are followed; a culture medium or diet containing all the basic requirements of the organism except one is supplied, the missing factor is then given in measured amounts and the growth response of the organism to these measured additions is noted. A graph is then plotted showing the variation of growth response with added growth factor—unknown samples can then be interpolated in the curve. In some investigations animals are used and their growth response is estimated by increase in weight, but it is now more usual to employ micro-organisms because they are both cheaper and easier to control. Growth response in micro-organisms is measured in terms of increased turbidity, cellular nitrogen, numbers of cells, or indirectly by measuring a metabolic product such as carbon dioxide.

Animal assay is used widely to test the effects of new therapeutic substances. Mice, rats, hamsters, guinea-pigs, monkeys, dogs, cats, swine and cattle are all used, as are birds, chickens, frogs, reptiles and insects. Tests are made especially in the fields of pharmacology and toxicology to estimate the therapeutic value of new drugs and the poisonous effects of potentially harmful substances. The tests do not always require whole animals but can sometimes be carried out on isolated parts—strips of tissue or single organs maintained in living, reproductive conditions. Animals and animal tissues have been widely used in the fields of virology and in the study of cancer. Frequently it is only by observing the response produced in a living animal that realistic measurements of the true effects of medical products can be obtained. Preparations of hormones, vaccines, anti-toxins, immune serums for example, must be tested in this way. The potency of digitalis extracts can be measured by injecting them into pigeons and measuring the amounts needed to cause the heart to stop.

Since different individual animals yield variations in their response to the same preparation, it is necessary to use statistical methods to interpret the results of a series of such tests. In fact both the test itself and the results must be treated statistically in order to obtain a reliable measure of the effects of the drug or toxin in terms of potency. The use of statistics in this study is known as biometrics.

The growth of knowledge about microbial nutrition between 1936 and 1945 stimulated the development of microbiological assay methods as a routine tool to test vitamins and amino-acids in natural substances. Such tests have the necessary precision, speed and sensitivity for detecting the presence and following the isolation of new vitamins and vitamin-like substances present in trace amounts only. Nicotinic acid, pantothenic acid, inositol, biotin, pyridoxal, etc., discovered by microbiological assay were only later found to be important elements in animal nutrition. Folic acid and vitamin B_{12} were also isolated by microbiological assay and in some cases this is the only method for the rapid and specific assay of individual vitamins. There are several advantages over animal assay. It requires less materials, time and labour and is therefore cheaper to perform than animal assay. It is also generally more precise but micro-organisms require pure chemicals if the results of the tests are to be interpreted correctly and this has meant the separation of vitamins from combinations in which they generally occur. This is a disadvantage of the method although it has led to the recognition, isolation and quantitative determination of certain previously unknown combined forms among the vitamins.

Micro-organisms are known which require one or more of all the water-soluble vitamins except ascorbic acid. Yeasts and lactic acid bacteria have been most widely used to assay these substances. Amino-acids can also be assayed by the use of micro-organisms, each of which has specific growth requirements for individual amino-acids. The basal media used contain a complete assortment of vitamins and the appropriate mixture of amino-acids except the one to be determined. Measured doses of this are then added.

Carbohydrate metabolism and the Krebs cycle

The term metabolism covers all the chemical changes which digested food molecules undergo *after* they have been absorbed into the body tissues. Studies of metabolic pathways have been made using heavy or radioactive isotopes to label specific compounds. The fate of the labelled atoms in the body can then be traced. The information is not direct and may therefore be misleading but much has been learned about metabolic pathways in the tissues. Proteins, lipids, carbohydrates and other constituents of the diet all have metabolic pathways which are cross-linked at many points. Here we shall discuss only the carbohydrates because they lead most readily into the well-known and extremely important biochemical principle—the Krebs cycle.

During digestion carbohydrates are first hydrolysed to mono-saccharides which are soluble in water and can be absorbed through intestinal mucosa into the blood-stream. The main carbohydrates involved are starch, glucose, sucrose and lactose. These produce glucose, fructose and galactose as hydrolysis products which are conveyed by the portal circulation to the liver where they are converted mainly into glucose. The main function of carbohydrates in the body is to provide a source of energy for biological oxidations, but they are also stored as glycogen which is found in glycosides and is converted into lipids or proteins.

Approximately half the daily energy requirements for a man is stored as glycogen in the liver (110 gm.) and the muscles (250 gm.). Liver glycogen is reversibly converted into glucose which finds its way into the blood-stream, whilst muscle glycogen is irreversibly converted into lactic acid.

The glucose level in human blood is fairly constant for a given individual, although it varies quite widely from one person to another. The rates at which carbohydrates are oxidised or converted into fats govern the concentration of blood glucose, whilst the hormones produced by the pancreas, adrenals, pituitary and thyroid play an important part in controlling the blood sugar level.

In the body glucose is oxidised to lactic acid in a number of small steps each of which involves the action of an enzyme. These processes occur largely in the muscles and when energy is liberated which is not immediately required by the muscle it is stored in energy-rich compounds including the adenosine phosphates.

Adenosine Triphosphate (ATP)

When this compound is required to furnish energy it is hydrolysed and a phosphate group is split off to leave adenosine diphosphate (ADP). At the same time about 10K cal per molecule is liberated. ATP can be regenerated from ADP by reaction with phosphate together with an energy boost to push it back up the scale.

$$\begin{array}{c} +H_2O \\ ATP \rightleftharpoons ADP + H_3PO_4 + 10K \text{ cal.} \\ -H_2O \end{array}$$

Another energy-rich compound is phosphocreatine, which like ATP can be hydrolysed to form creatine and phosphoric acid with the liberation of energy, thus

$$
\begin{array}{ccc}
\underset{\substack{| \\ \text{COOH}}}{\overset{\substack{\text{CH}_3 \\ |}}{\text{CH}_2}}\text{—N—}\overset{\substack{\text{H} \\ |}}{\underset{\substack{\| \\ \text{NH}}}{\text{C}}}\text{—}\overset{\substack{\text{O} \\ \|}}{\underset{\substack{| \\ \text{OH}}}{\text{N—P}}}\text{—OH}
& \overset{+\text{H}_2\text{O}}{\underset{-\text{H}_2\text{O}}{\rightleftharpoons}}
& \underset{\substack{| \\ \text{COOH}}}{\overset{\substack{\text{CH}_3 \\ |}}{\text{CH}_2}}\text{—N—}\overset{}{\underset{\substack{\| \\ \text{NH}}}{\text{C}}}\text{—NH}_2 + \text{H}_3\,\text{PO}_4 + 10\text{K cal}
\end{array}
$$

Phosphocreatine   *Creatine*

Some of the chemical reactions involved in carbohydrate metabolism and muscular contraction are related with these energy-rich phosphates.

It has been stated that in the muscles glucose is converted into lactic acid with the evolution of energy, but the complete oxidation of glucose to carbon dioxide and water would yield very much more energy. From this we may conclude that large amounts of energy remain in the lactic acid molecule. Now lactic acid does not accumulate in the muscles; about 25 per cent of it is oxidised to carbon dioxide and water and the rest is converted into glycogen. The oxidation of lactic acid is brought about by inspired oxygen carried by the blood-stream and the large amounts of energy liberated in the process are used to convert ADP into ATP. The energy so absorbed is then evolved again in the conversion of the greater part of the lactic acid into glycogen, partly in the muscles but mainly in the liver. The more work we do the more lactic acid is formed and the increased rate of breakdown of glycogen to lactic acid results in increased ATP formation. The rate at which these changes can occur is limited by the rate of respiration. Prolonged strenuous exercise produces fatigue when the lactic acid formed exceeds the rate at which the body can supply oxygen to oxidise it.

Table XVII KREBS CYCLE

KREBS CYCLE

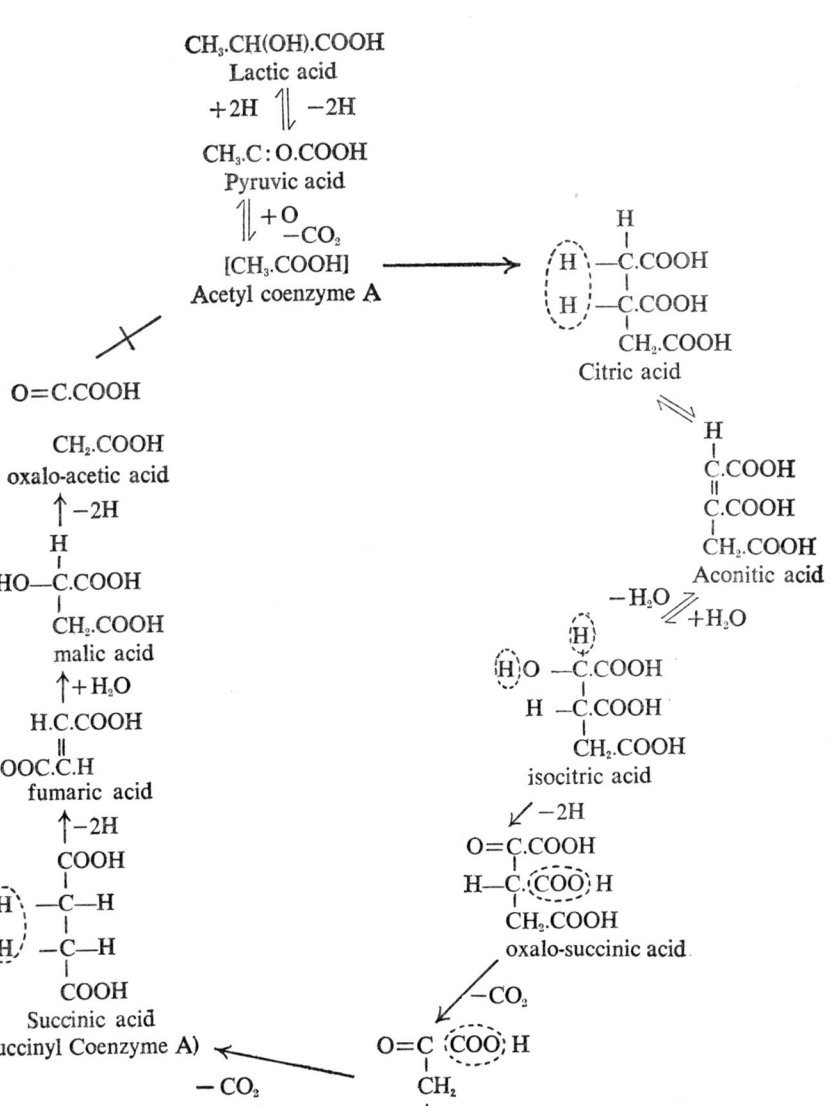

CH₃.CH(OH).COOH
Lactic acid
+2H -2H

CH₃.C:O.COOH
Pyruvic acid
+O -CO₂

[CH₃.COOH]
Acetyl coenzyme A

H
|
H—C.COOH
|
H—C.COOH
|
CH₂.COOH
Citric acid

O=C.COOH
|
CH₂.COOH
oxalo-acetic acid
-2H

H
|
HO—C.COOH
|
CH₂.COOH
malic acid
+H₂O

H.C.COOH
‖
HOOC.C.H
fumaric acid
-2H

COOH
|
H—C—H
|
H—C—H
|
COOH
Succinic acid
Succinyl Coenzyme A)

H
|
C.COOH
‖
C.COOH
|
CH₂.COOH
Aconitic acid
-H₂O +H₂O

H
|
HO—C.COOH
|
H—C.COOH
|
CH₂.COOH
isocitric acid
-2H

O=C.COOH
|
H—C.(COO)H
|
CH₂.COOH
oxalo-succinic acid
-CO₂

O=C (COO)H
|
CH₂
|
CH₂.COOH
α Keto-glutaric acid

-CO₂
[+O]

The complex series of reactions by which lactic acid is oxidised in the body was first proposed in 1937 by Sir Hans Krebs, who was at that time working at Sheffield University. It is often called the citric acid or tricarboxylic acid cycle. Lactic acid is first oxidised to pyruvic acid ($CH_3.C:O.COOH$) which is then changed by the aid of coenzyme A into citric acid with the evolution of a molecule of carbon dioxide.

Coenzyme A

Citric acid is then successively converted through a complex series of organic acids in the course of which two more molecules of carbon dioxide are released. The overall process can be represented by the equation,

$$CH_3.CHOH.COOH + 3O_2 \rightarrow 3CO_2 + 3H_2O + Energy.$$
lactic acid

A great deal of detail has been worked out in the study of these processes, but very little is yet known about the way in which certain carbon atoms are moved from one compound to another within the cycle. In one direction the chain of reactions yields energy, whilst in the other it forms a mechanism for the synthesis of essential constituents. It also functions as a means of shifting metabolism from one major pathway to another—for example from the metabolism of carbohydrates to that of fats. Physiological studies indicate that such changes do in fact occur, but it is not yet known how the cycle 'decides' which way it must function at each instant. Nevertheless this cycle is known to be the key by which the regulation of metabolic pathways is controlled.

Conclusion

Complexity appears to be the keynote of biochemistry—complexity both in chemical constitution and in the sequence of chemical changes which are involved in natural processes. Much of this complexity is caused by the juxtaposition of many smaller simpler units, as of amino-acids in the proteins or the limited steps of oxidation in the Krebs cycle. Such arrangements are eminently suited to analysis by the computer and we can expect to see an increase in the use of this instrument in the future. Indeed, sophistication of procedure has been essential in the solution of biochemical problems and there is every indication that this must be increased still further if we are to unravel all the finer strands of the story. In this book we have dealt with only about half the picture in any case, for biochemistry is quite as much concerned with the chemical constitution and functions of plants as with those of animals and photosynthesis, the primary life process, has caused at least as many problems as metabolism. Modern biochemistry has derived as much from the rise of vegetable chemistry as from that of animal chemistry and the two aspects of the subject are found to have much in common.

So far the extension of understanding has been the principal objective of biochemistry, although the animal chemists in general had a practical aim in view since they sought to improve medical treatment. Until recently biochemistry has remained in this first stage of development and the attempt to duplicate life processes *in vitro* has played only a minor part in the activities of the biochemist. But it is always the intention of scientists so to understand nature that her activities can be both controlled and copied. Biochemistry is now moving into this second stage of development. Microbiological techniques, manipulation of genes, cloning, etc., are all under investigation. At present success has been achieved only with simple organisms but ultimately we can expect them to be applied to the higher animals including man himself. Here the biochemist is treading upon dangerous ground for he is entering regions of moral and philosophical arguments concerning the nature of man and his relationship with the rest of the natural world. It is not impossible that the biochemist may find, as the physicist has already found, that he is dealing with forces more powerful than he suspected. His knowledge may then be coveted by ruthless men who see in it the means of controlling completely the lives of vast numbers of the human race. In the end we shall find that the complete understanding of life requires not only great experimental ingenuity but also such human values as integrity, tolerance and moral courage.

The ultimate question to which the biochemist may yet be able to supply an answer is concerned with the origins of life itself. How did life on Earth begin? For a long time man has accepted the concept of a Creator and even after Darwin's theory of evolution by natural selection had shaken the common faith in the uniqueness of Man, the strength of religious beliefs restored faith in the Divine Creation. Yet, with his reservations about the concept of vital force, the animal chemist took the first steps along the road to a rational explanation of the mechanisms of life. It seems that the phenomena of life are produced by a finely balanced, complex system of physical and chemical changes. It is from the very complexity of the living organism that we are led to suspect that there is something more than the merely physical present. But if we reject special creation we are left with one alternative only— life must have arisen at some appropriate time in the history of the Earth from the materials present on its surface. The biochemist is now in a good position to study this problem with his intimate knowledge of cell constituents, proteins and nucleic acids, cell metabolism and the essential components of the environment to perpetuate the life processes.

The origins of life can now be seen in relationship to the long story of evolution, beginning with the interconversion between energy and elementary matter, passing through the formulation of nebulae, stars and planets. The whole process is driven by energy released in nuclear fusion which is also the source of energy responsible for maintaining life itself. There are so many stars in the Universe that it is almost certain that there are planets like the Earth circling some of them and it is so likely that other forms of life exist on some of these planets as to be almost a certainty. Man can no longer think that he is in any way unique, though whether it will ever prove possible for us to communicate with other intelligent forms of life elsewhere in the Universe remains an open question.

In these ways the study of biochemistry begins to find links with physics and cosmology. It has expanded man's vision and helped to change his ideas about himself and about his relationship to the rest of the Universe. Precisely where these ideas may lead us remains to be seen but it cannot be doubted that biochemistry is a fundamental study of crucial importance to us all.

Bibliography

A. Further reading in the history of animal chemistry and related topics.

Multhauf, R. P. *The Origins of Chemistry*, Oldbourne, London, 1966.

Debus, A. *The English Paracelsians*, London, 1965.

Goodfield, G. J. *The Growth of Scientific Physiology*, London, 1960.

Mendelsohn, E. *Heat and Life; the development of the Theory of Animal Heat*, Harvard U.P., 1964.

McCollum, E. V. *A History of Nutrition*, Boston, U.S.A., 1957.

Gottlieb, Leon S. *A History of Respiration*, Springfield, U.S.A., 1964.

Smeaton, W. A. *Fourcroy; Chemist and Revolutionary, 1755-1809*, Cambridge, 1962.

Crosland, M. *The Society of Arcueil, a view of French science at the time of Napoleon I*, Heinemann, London, 1967.

Olmsted, J. M. D. and Olmsted, E. H. *Claude Bernard and the Experimental method in Medicine*, New York, 1944.

Costa, Albert B. *Michel Eugène Chevreul, 1766-1889, Pioneer of Organic Chemistry*, Univ. of Wisconsin Press, 1962.

Drabkin, D. L. *Thudichum, Chemist of the Brain*, Univ. of Pennsylvania Press, 1958.

Nicolle, J. *Louis Pasteur, A master of scientific enquiry*, London, 1961.

Plimmer, R. H. A. *The History of the Biochemical Society*, Cambridge, 1949.

Stuyvenberg, J. H. van (editor), *Margarine; An Economic, Social and Scientific History*, Liverpool U.P., 1969.

Florkin, M. *A History of Biochemistry*, 4 Vols, (Section VI, Vols 30-33 of *Comprehensive Biochemistry*), Elsevier Vol 30, 1972 (Vols 31-33 in press).

B. Some sources and specialised works in the history of animal chemistry and early biochemistry.

Hales, S. *Statical Essays; containing Haemastatics; Or an Account of some Hydraulick and Hydrostatical Experiments made on the Blood and Blood-vessels of Animals. Also, an Account of some Experiments on Stones in the Kidneys and Bladder; with an Enquiry into the Nature of these anomalous Concretions.* London, 1733 (second ed. 1740, third ed. 1769). Repr. Hafner, London, 1964.

Crawford, A. *Experiments and Observations on Animal Heat and the Inflammation of Combustible Bodies.* London, 1779, (second ed. 1788).

Fourcroy, A. F. de, *Système des Connaissances Chimiques, et leurs Applications aux Phénomènes de la Nature et de l'Art.* 11 Vols, Paris, 1801-2, (English trans, W. Nicholson London, 1804).

Johnson, W. B. *History of the Progress and Present State of Animal Chemistry*, 3 Vols, London, 1803.

Bostock, J. *An Essay on Respiration*, 2 parts, Liverpool and London, 1804.

Allen, W. and Pepys, W. H. On the Changes produced in Atmospheric Air, and Oxygen gas, by Respiration, *Phil. Trans, 98*, 242-267, 1808; *Ibid., 99*, 404-429, 1809.

Berzelius, J. J. *A View of the Progress and Present State of Animal Chemistry*, trans, G. Brunmark, London, 1813, (second ed. 1819).

General Views of the composition of Animal Fluids, *Medchir. Trans, 3*, 198-276, 1812; *Ann. Phil, 2*, 19-26, 195-208, 377-387, 415-425, 1813.

Lehrbuch der Chemie, trans, F. Wöhler, 4 Vols, Dresden, 1825-31, (third ed. trans, F. Wöhler, 10 Vols, Dresden and Leipzig, 1833-41, fourth ed. trans, F. Wöhler, 10 Vols, Dresden and Leipzig, 1835-41, fifth ed. 5 Vols, Dresden and

Leipzig, 1843-8). French trans, Jourdan and Esslinger, *Traité de Chimie*, 8 Vols, Paris, 1829-33.

Prout, W. *An Inquiry into the Nature and Treatment of Gravel, Calculus and other Diseases connected with a Deranged Operation of the Urinary Organs*, London, 1821, (second ed.) rev. and enlarged, London, 1825, third ed. *On the Nature and Treatment of Stomach and Urinary Diseases*, etc., London, 1840, (fourth ed. 1843, fifth ed. 1848).

Chemistry, Meteorology and the Function of Digestion considered with Reference to Natural Theology, (eighth Bridgewater Treatise), 2 eds. London, 1834.

Observations on the quantity of Carbonic Acid gas emitted from the lungs during respiration, at different times and under different circumstances, *Ann. Phil, 2*, 328-343, 1813; *Ibid., 4*, 331-7, 1814.

On the Phenomena of Sanguification and on the Blood in general, *Ann. Phil, 13*, 12-25, 268-279, 1819.

On the Nature of the acid and saline matters usually existing in the stomachs of Animals, *Phil. Trans, 114*, 45-49, 1824.

Observations on the Nature of some of the Proximate Principles of the Urine, *Med-chir. Trans, 8*, 526-549, 1817; *Ibid., 9*, 472-484, 1818.

Marcet, A. *An Essay on the Chemical History and Medical Treatment of Calculous Disorders*, London, 1817, (second ed. 1819).

Philip, A. P. W. *An experimental inquiry into the laws of the Vital Functions, with some Observations on the Nature and Treatment of Internal Diseases*, London, 1818.

Chevreul, M. E. *Recherches Chimiques sur les Corps Gras D'origine Animale*, Paris, 1823.

Considérations Générales sur l'analyse organique et sur ces Applications, Paris, 1824.

Leuret, F. and Lassaigne, J. L. *Recherches Physiologiques et*

Chimiques pour servir à l'Histoire de la Digestion, Paris, 1825.

Tiedmann, F. and Gmelin, L. *Recherches Expérimentales Physiologiques et Chimiques sur la Digestion Considerée dans les Quatre Classes d'Animaux Vertébrés*, trans, A. J. L. Jourdan, 2 Vols, Paris, 1827.

Macaire, I. F. and Marcet, F. Recherches sur l'origine de l'Azote qu'on retrouve dans la composition des Substances Animales, *Ann. Chim*, (2), *51*, 371-395, 1832.

Boussingault, J. B. Recherches sur la quantité d'Azote contenue dans des Fourrages, et sur leurs Equivalens, *Ann. Chim*, (2), *63*, 225-244, (1836).

Analyses comparées des Alimens consommés et des Produits rendus par une vache laitière; recherches entreprises dans le but d'examiner si les animaux herbivores empruntent de l'azote à l'atmosphère, *Ann. Chim.* (2), *71*, 113-136, 1839; *Ibid.*, (3), *11*, 433-456, 1844.

Beaumont, W. *Experiments and Observations on the Gastric Juice and the Physiology of Digestion*, Plattburg, U.S.A., 1833. (Repr. Dover, London, 1959).

Edwards, W. F. *De l'influence des agens physiques sur la vie*, Paris, 1824. English trans, Hodgkin and Fisher, London, 1832.

Liebig, J. von, *Animal Chemistry, or Organic Chemistry in its Application to Physiology and Pathology*, ed. W. Gregory, London, 1842. (Repr, with intro. by F. L. Holmes, New York, 1964.)

Jones, H. Bence *On Gravel, Calculus and Gout; chiefly an Application of Professor Liebig's Physiology to the Prevention and Cure of these Diseases*, London, 1842.

On Animal Chemistry in its Applications to Stomach and Renal Diseases, London, 1850.

Lectures on some of the Applications of Chemistry and Mechanics to Pathology and Therapeutics, London, 1867.

Thomson, T. *The Chemistry of Animal Bodies*, Edin, 1843.

Scharling, E. A. Recherches sur la quantité d'acide carbonique Expiré par l'Homme dans les vingt-quatre heures, *Ann. Chim*, (3), *8*, 478-497, 1843.

Bird, G. *Urinary Deposits, their Diagnosis, Pathology and Therapeutical Indications*, London, 1844.

Lectures on Electricity and Galvanism in their Physiological and Therapeutical Relations, London, 1849.

Matteucci, C. *Lectures on the Physical Phenomena of Living Beings*, trans, J. Pereira, London, 1847.

Laskowski, N. Ueber die Proteintheorie, *Liebig's Ann, 57*, 129-166, 1846.

Carpenter, W. B. On the Mutual Relations of the Vital and Physical Forces, *Phil. Trans, 140*, 727-757, 1850.

Dumas, J. B. A. and Boussingault, J. B. *Essai de Statique Chimique des Êtres Organisés*, Paris, 1841. English trans, (anon), *The chemical and Physiological Balance of Organic Nature*, London, 1844.

Bidder, F. and Schmidt, C. *Die Verdaungssaefte und der Stoffwechsel Mitau*, 1852.

Magendie, F. et al, Rapport fait à l'Académie des Sciences au nom de la Commission de la Gélatine, *Compt. Rend. Acad. Sci*, Paris, 13, 237-283, 1841.

Lehmann, C. G. *Physiological Chemistry*, trans, Geo. E. Day, 3 Vols, London, 1851-4.

Simon, J. F. *Animal Chemistry with reference to the Physiology and Pathology of Man*, trans, Geo. E. Day, 2 Vols, London, 1865.

Thudichum, L. J. W. *A Treatise on gallstones and their chemistry, pathology and treatment*, London, 1863.

A Treatise on the Chemical Constitution of the Brain, London, 1884.

Starling, E. H., The Croonian Lectures on the chemical correlation of the functions of the body, *Lancet*, 1905, (ii), 339-341, 423-5, 501-3, 579-583.

Hofmann, A. W. The Lifework of Liebig in Experimental and Philosophical Chemistry, etc., *J. Chem. Soc, 28*, 1065-1875; Repr. *Chemical Society, Faraday Lectures*, London, 1928, pp. 44-119.

Young, F. G. Claude Bernard and the Theory of the Gllycogenic Function of the Liver, *Ann. Sci, 2*, 47-83, 1937. Biochemistry of the endocrine system, *Adv. of Sci, 21*, 369-378, 1964.

Harris, L. J. *Vitamins; a digest of current knowledge*, London, 1951.

Kasich, A. M. William Prout and the discovery of the hydrochloric acid in the gastric juice, *Bull. Hist. Med, 20*, 340-358, 1946.

Farber, E. Variants of Preformation Theory in the History of Chemistry, *Isis, 54*, 443-460, 1963.

Holmes, F. L. Elementary Analysis and the Origins of Physiological Chemistry, *Isis, 54*, 50-81, 1963.

Lipmann, T. O. Wöhler's Preparation of Urea and the Fate of Vitalism, *J. Chem. Educ, 41*, 452-458, 1964; Vitalism and Reductionism in Liebig's Physiological Thought, *Isis, 58*, 167-185, 1967.

Brooke, J. H. Wöhler's Urea, and its vital force?—a verdict from the chemists, *Ambix, 15*, 84-114, 1968.

Teich, M. On the Historical Foundations of Modern Biochemistry, *Clio. Medica, 1*, 41-57, 1968.

Partington, J. R. *A History of Chemistry*, 4 Vols, London, 1961.

Keilin, D. (ed. Joan Keilin), *The History of Cell Respiration and Cytochrome*, Camb. Univ. Press, 1966.

Needham, Dorothy M. *Machina Carnis, The Biochemistry of Muscular Contraction in its Historical Development*, Camb. Univ. Press, 1972.

Systematic Chemical Nomenclature

Organic compounds

Common Name	*Systematic Name*
acetaldehyde	ethanal
acetylene	ethyne
acids:	acids:
citric	2-hydroxy propane-1, 2, 3-tricarboxylic
lactic	2-hydroxy propanoic
malic	2-hydroxy butanedioic
oxalic	ethane dioic
pyruvic	2-oxopropanoic
sebacic	decanedioic
stearic	octadecanoic
succinic	butane dioic
tartaric	2, 3-dihydroxybutane dioic
acrolein	propenal
alcohol (ethyl)	ethanol
catechol	benzene-1, 2 diol
cetyl alcohol	hexadecan-1-ol
ethylene dibromide	1, 2-dibromo ethane
glycerine/glycerol	propane-1, 2, 3-triol
glycine	amino acetic acid
pyrogallol	benzene-1, 2, 3-triol
quinone	cyclohexadiene-1, 4-dione
vinyl	ethenyl

Inorganic compounds

Common Name	*Systematic Name*
baryta	barium (II) hydroxide
chlorate ion	chlorate (v) ion
copper oxide	copper (II) oxide
ferrocyanide ion	hexacyanoferrate (II) ion
lead dioxide	lead (IV) oxide
limewater	calcium (II) hydroxide solution
magnesia	magnesium (II) oxide
nitrate ion	nitrate (V) ion
phosphate ion	phosphate (V) ion
sulphate ion	sulphate (VI) ion

Index